T0203682

Correction of Differential Settlements in Mexico City's Metropolitan Cathedral and Sagrario Church

Built Heritage and Geotechnics
Series editor: Renato Lancellotta

Volume II

ISSN: 2640–026X
eISSN: 2640–0278

Correction of Differential Settlements in Mexico City's Metropolitan Cathedral and Sagrario Church

Efraín Ovando-Shelley
Instituto de Ingeniería, Universidad Nacional,
Autónoma de México, Mexico City, Mexico

Enrique Santoyo†
TGC Geotecnia, Geotechnical Consultants, Mexico City, Mexico

CRC Press
Taylor & Francis Group
Boca Raton London New York Leiden

CRC Press is an imprint of the
Taylor & Francis Group, an **informa** business

A BALKEMA BOOK

CRC Press/Balkema is an imprint of the Taylor & Francis Group, an informa business

© 2020 Taylor & Francis Group, London, UK

Typeset by Apex CoVantage, LLC

Library of Congress Cataloging-in-Publication Data
Names: Ovando Shelley, Efraín, author. | Santoyo, Enrique (Santoyo Villa), author.
Title: Correction of differential settlements in Mexico City's Metropolitan Cathedral and Sagrario Church / Efraín Ovando-Shelley, Instituto de Ingeniería, Universidad Nacional, Autónoma de México, Mexico City, Mexico, Enrique Santoyo, TGC Geotecnia, Geotechnical Consultants, Mexico City, Mexico.
Description: First edition. | Boca Raton : CRC Press/Taylor & Francis Group, [2020] | Series: Built heritage and geotechnics, 2640026X ; vol 2 Includes bibliographical references and index.
Identifiers: LCCN 2019051006 (print) | LCCN 2019051007 (ebook) | ISBN 9780367344887 (hardback) | ISBN 9780429326219 (ebook)
Subjects: LCSH: Catedral de México. | Parroquia del Sagrario Metropolitano (Mexico City, Mexico) | Church architecture—Conservation and restoration—Mexico—Mexico City. | Architecture, Spanish colonial—Conservation and restoration—Mexico—Mexico City. | Cathedrals—Conservation and restoration—Mexico—Mexico City. | Earth movements and building—Mexico—Mexico City. | Soil stabilization.
Classification: LCC NA5257.M4 O93 2020 (print) | LCC NA5257.M4 (ebook) | DDC 726.50972/53—dc23
LC record available at https://lccn.loc.gov/2019051006
LC ebook record available at https://lccn.loc.gov/2019051007

Published by: CRC Press/Balkema
Schipholweg 107C, 2316 XC Leiden, The Netherlands
e-mail: Pub.NL@taylorandfrancis.com
www.crcpress.com – www.taylorandfrancis.com

ISBN: 978-0-367-34488-7 (Hbk)
ISBN: 978-0-429-32621-9 (eBook)
DOI: https://doi.org/10.1201/9780429326219

Contents

Foreword

This is the second of a series of volumes on Built Heritage and Geotechnics, intended to reach a wide audience: professionals and academics in the fields of civil engineering, architecture and cultural resources management, and particularly those involved in art history, history of architecture, geotechnical engineering, structural engineering, archaeology, restoration and cultural heritage management, and even the wider public.

Motivations of this series rely on the fact that preservation of built heritage is one of the most challenging problems facing modern civilization. It involves in inextricable patterns various cultural, humanistic, social, technical and economic aspects. The complexity of the topic is such that a shared framework of reference is still lacking among art historians, architects, structural and geotechnical engineers. This is proved by the fact that, although there are exemplary cases of an integral saving of any structural components with its static and architectural function, as a material witness of the knowledge, the culture and the construction techniques of the original historical period, there are still examples of uncritical confidence in modern technology which leads to the replacement of previous structures with new ones, which only preserve an iconic appearance of the original monument.

For these reasons, publishing short books on specialized topics – such as well-documented case studies of restoration work at one specific site or of a monument; detailed overviews of construction techniques intended as a material witness of knowledge of the historical period in which the monuments were built; or specific conservation/preservation works – may be of great value.

The present volume, titled *Correction of Differential Settlements in Mexico City's Metropolitan Cathedral and Sagrario Church*, provides essential information on the history of the construction, the involved techniques and interventions applied over time to mitigate the effects of differential settlements and finally describes a rather unique solution that in the technical literature is today known as *underexcavation*.

The Metropolitan Cathedral in Mexico City is one of the most important and spectacular examples of architectural heritage in the Americas. The pronounced differential settlements suffered by the Cathedral and the adjacent Sagrario Parish Church since the earliest stages of their construction originated from the fact that they were built on extremely soft lacustrine clays over the remains of Aztec temples. The properties of these lacustrine clays are rather singular: the water content may be as high as 300% and the compressibility may be an order of magnitude higher than those values geotechnical engineers are familiar with, to mention just a few properties.

In addition, soils are to be considered "*materials with memory*", which means that their mechanical behaviour is the heritage of the so-called stress history, stretching from the early deposition phase to the most recent loading processes. The reader may also find this relevant concept in the companion book of this series devoted to the Cathedral of Modena and the Ghirlandina Tower and explore many similarities regarding the causes of differential settlements.

In the case under examination, loading imposed on the soil by Aztec temples and subsequent unloading when they were destroyed by Cortés in 1521 gave rise to a loading history, with the consequence of inducing spatial variation of compressibility in the foundation soil and therefore differential settlements.

In spite of a number of underpinning attempts undertaken in the past, differential settlements continued to increase, and, in recent years, at an alarming rate, as a consequence of the regional subsidence, induced by the exploitation of deep aquifers.

It could be expected that a uniform drop of groundwater level in the region should not induce differential settlements. But this was not the case, and the reader will discover that the uniform drop of groundwater level generated differential settlements because of non-homogeneous compressibility of the soft clay horizon, linked to the mentioned loading and unloading history induced by the Aztec temples, as well as the interaction between neighbouring structures.

Therefore, to preserve the Cathedral and the Sagrario Church, it was decided to apply the method of *underexcavation*: soil was removed at selected locations below the foundations in order to induce settlements to compensate those previously induced by loads and subsidence.

This aspect is really a fascinating one, because this idea was first suggested by Leonardo Terracina, as published in *Géotechnique* in 1962, to correct the leaning of the Tower of Pisa, and it was successfully applied to the Tower of Pisa after the experience gained at Mexico City.

In short, by considering the mentioned peculiarities of this case history, the reader will certainly be encouraged to deeply scrutinize this book as well as the forthcoming one on the Tower of Pisa to explore similarities and differences in the application of this technique of underexcavation, that in both cases proved to have the benefit of preserving the full integrity of the monument.

The series editor
Renato Lancellotta

Preface

In 1989, underexcavating the Metropolitan Cathedral in Mexico City appeared to be a bold and daring idea that, locally, was met with scepticism by the authorities and was acrimoniously criticized by influential geotechnicians. The idea, however, did not come up as a result of a spontaneous occurrence. Rather, it was put forth after identifying the originating cause of damage accumulated over several centuries and by considering possible solutions within the field of geotechnics. Indeed, a necessary first step towards restoring and preserving a monument is to identify the factor or factors that bring about damage. In Mexico City, extremely soft and compressible soils as well as regional subsidence in the former lakebed combine, giving rise to differential settlements in the city producing damages that gradually accumulate in the totality of the urban infrastructure, which can be especially detrimental on the city's architectural heritage. Further, since the city is also exposed to seismic hazard, buildings gradually become more vulnerable to earthquakes as differential settlements become larger.

Well-documented cases of underexcavation in the city were first reported by Enrique Santoyo (volume co-author) when he applied the method to correct the inclination of a few buildings that suffered sudden settlements and that tilted during the huge earthquake of 19 September 1985. It was he who assembled these cases to illustrate the possibilities of underexcavation as a potential solution to eliminate or mitigate differential settlements in the Cathedral and prepared the initial proposal to underexcavate it. The proposal was backed by Professor Fernando López-Carmona, from the School of Architecture (National University of Mexico). López-Carmona and Santoyo were then joined by Enrique Tamez, and the proposal was given due consideration by the authorities, most notably by Sergio Zaldívar, who assumed the responsibility of heading the project and for finding funds to carry it out.

Before the actual underexcavation at the Cathedral, a large-sized trial was carried out at the Church of San Antonio Abad (Saint Anthony the Abbot), as described in the body of this text. Results were very promising, but still, it was decided to discuss these and to evaluate the project that had been developed after the trial. The technical staff at that time now included Professor Roberto Meli and Roberto Sánchez, both structural engineers from the Instituto de Ingeniería (National University of Mexico) and Efraín Ovando-Shelley (this volume's first author), who joined Santoyo and Tamez as part of the geotechnical team.

An international committee was formed to evaluate the project. This committee was formed by geotechnical and structural engineers, some of whom also formed part of the committee for the safeguarding of the Tower of Pisa, and by local engineers. The geotechnical consultants that formed part of the committee were Professors Gholamreza Mesri, Michele Jamiolkowski and John B. Burland; local geotechnical experts included Professors Daniel Reséndiz, Gabriel Auvinet and Miguel Romo. The committee met in November 1992 and examined the project thoroughly and comprehensively for three days, after which the project was given the go-ahead. Underexcavation went on from July 1993 to June 1998.

Santoyo also sought and found a means of preserving the corrective settlements achieved with underexcavation when the soil removal operations finished. Selective soil hardening by injecting setting fluids was then applied from 1999 to 2001. As shown here, the procedure, designed to reduce or mitigate the apparition of differential settlements, has proven to be successful.

This text provides a brief account of the intervention of more than 10 years in which the technical staff headed by Santoyo worked and learned with much more than simple enthusiasm. It has been highly gratifying to see through instrumental observations that, indeed, the intervention had a beneficial effect on both the Cathedral and the Sagrario Church, as can be demonstrated with data gathered over the years.

Enrique Santoyo died in June 2016 at a time when, thankfully, he was able to ascertain that underexcavation and soil hardening in the Metropolitan had been successful. The original version of the text presented here is based on a version in which Enrique and Ovando-Shelley worked on together, in Spanish. The final draft is an update of that original version, including data gathered later. Ovando-Shelley produced the final draft in English.

Finally, all the plots and all of the graphic material were produced by TGC Geotecnia. The help of Julio César Hernández is duly acknowledged, as well as the contribution of Hugo Ortega in adapting that material to the needs of this text. Santoyo's son, Enrique Santoyo-Reyes, also enthusiastically supported the production of this book.

Chapter 1

Introduction

According to official historical records, the Aztecs founded their capital, later Mexico City, in 1325 on an islet in the middle of Lake Texcoco, the lowest portion of a closed basin surrounded by volcanoes. Prevalence of a lacustrine environment during long periods propitiated the deposition of large volumes of fly ash and other pyroclastic materials that were driven into the lake by water flows running from the higher lands and by the winds. In time, the chemical degradation of these materials formed clays and clayey silts. These soils are geologically very young and are notorious for their extremely large water contents and compressibility. Hence, colonial buildings of the 16th, 17th and 18th centuries have undergone large differential settlements due to self-weight consolidation. Towards the middle of the 19th century, pumping from deep wells induced regional subsidence in the city and additional differential settlements were brought about on the old buildings.

Self-weight consolidation and regional subsidence have exerted their influence simultaneously on any building erected in Mexico City after deep well pumping began, and that is why it has become increasingly necessary in Mexico City to correct the behaviour of buildings that have been damaged by differential settlements. It is not always possible to apply the conventional solutions provided by modern foundation engineering to solve these problems, either because they are too expensive or because in applying them, the original structure must be altered, a situation which is usually not allowed in historic buildings. Less obtrusive solutions have been used recently to correct the behaviour of many historic buildings around the world. One of these, the method of underexcavation, has been applied successfully in Mexico City on a number of buildings to reduce their differential settlements and inclinations. Soil removed at selected locations below the foundations of structures produces settlements in the surface which will compensate those brought about by the consolidation of the soil mass due to permanent loads or by effective stress increments that result from deep well pumping or by earthquake-induced deformations. The technique was first presented formally by Terracina (1962) to reduce the inclination of the Tower of Pisa.

Underexcavation of buildings in Mexico City was first used in the early seventies. Well-documented projects and the rational and systematic application of the method began after the 1985 earthquakes, when many buildings lost their plumb and, in some cases, tilted dangerously. In 1989, damage that accumulated in the Metropolitan Cathedral prompted local authorities to initiate a project to salvage and preserve it and the adjoining Sagrario Church, which were underexcavated between August 1993 and June 1998. The project was officially named *Corrección geométrica de la Catedral y el Sagrario Metropolitanos de la Ciudad de México* (*Geometrical correction of the Metropolitan Cathedral and its adjacent Sagrario Church*). Its purpose was to reduce differential settlements that had accumulated mostly during the last century.

Several options were analyzed thoroughly, and after careful consideration, the method of underexcavation was finally chosen after a successful large-scale trial at a basilical church built some 1,500 m from the cathedral in a similar geotechnical environment, also built with similar materials and construction procedures. Underexcavation, a term coined in this project, consisted in drilling small-diameter tunnels through the underlying clays. Successive opening and closing of the holes induced progressive corrective settlements until the magnitude of the cumulative settlement, which varied from one point to the other, became compatible with the objective set forth. This levelling work was subsequently complemented with a preventive activity, the selective hardening of the subsoil, which decreased differential settlement rates.

In summary, the structural objectives were as follows: a) to sink the apse of the Cathedral from 80 to 95 cm, following a rigid body movement; b) to induce a closing or supplementary rotation of the lateral walls to strengthen the confinement of the "counteracting belt" constituted by the internal walls of the chapels and by those surrounding the church; and c) to apply a rotation to the Sagrario by generating a 30 cm subsidence of its north side. These objectives were accomplished, and the cumulative deformations of the last 65 years were eliminated. After the original goals of the project were fulfilled, underexcavation stopped, and therefore the religious complex was again exposed to the effect of the differential settlements brought about by regional subsidence.

Since the start of the project, the responsible parties were aware that the harmful effects of regional subsidence would reappear upon completion of the underexcavation works and that, as a result, underexcavation would have to be repeated periodically in the course of time. Towards the end of the corrective actions, several preventive options were analyzed. Two possibilities were contemplated: a) the use of pressure cement grouting and b) the construction of sand columns. The authors studied a most illuminating precedent, the case of the Palacio de Bellas Artes (Palace of Fine Arts), an opera house where mortar grouts were injected into the soil in 1907. Before applying this technique at the cathedral, an experimental program of clay hardening through field tests was first carried out at the former Texcoco Lake. Regarding the proportioning of the mortars used and the optimization of the drilling and grouting techniques, the experimental work applied to the installation of micropiles that had been performed in the last few years were also applied in this project. The success of the field tests justified the execution of an experimental section at the western atrium and the preparation of a project for soil hardening under the Cathedral and the Sagrario. The recommendation to carry out the full treatment ensued. Three annual stages of controlled grouting followed, applying the observational method to control it.

Geotechnical work was developed in five major stages: a) preliminary studies: b) detailed design and its experimental demonstration; c) execution of the underexcavation; d) experimental grouting; and e) mortar grouting of the soil mass.

Underpinning

The Cathedral and Sagrario suffered differential settlements during the 20th century in spite of two underpinning campaigns that were carried out to prevent their occurrence. The first one took place in the forties and it basically consisted in emptying the foundation cells; its effects were imperceptible because the causes of the phenomenon were not eliminated. During the second stage, performed in 1975, control piles were installed, but it was only possible to temporarily modify the behaviour of the foundation because the maximum load that can be supported by the piles is about 30% of the total weight of the Cathedral and, especially, because the control piles are not capable of eliminating the differential deformations of the subsoil. Mention should be made that the control piles under the Cathedral can only reduce the differential settlement rates; however, in the long term the total magnitude of the differential settlement will eventually be equal to that obtained without the use of such piles.

Acknowledgements

Our appreciation is extended to Sergio Zaldívar, head of the project until the year 2000, and to Dr. Xavier Cortés Rocha, who headed the project thereafter until 2006. The work of the technical committee members – Dr. Enrique Tamez González, Dr. Fernando López Carmona, Dr. Roberto Meli, Dr. Enrique Santoyo Villa and Mr. Hilario Prieto – is also duly acknowledged; Dr. Jorge Díaz Padilla acted as secretary of the committee. The advisory support group was formed by Dr. Efraín Ovando-Shelley, Mr. Roberto Sánchez, Mr. Alfredo Velasco and Mr. Arturo Ramírez Abraham.

Chapter 2

Description of foundations, general structural characteristics

The Metropolitan Cathedral was built on part of the land covered originally by the Aztec Ceremonial Precinct. Remains of some of the Aztec structures corresponding to this important pre-Hispanic archaeological site can still be seen under its foundation (Figure 2.1).

It is very likely that the foundations of the Metropolitan Cathedral may have been originally conceived with continuous footings running below walls and with individual footings under column axes, as was customary in Mexico City during the first few decades after the fall of the Aztec empire. The construction of these foundations began around 1560 and advanced slowly and painstakingly. Spanish masons and Aztec workers could barely walk on the marshy mud, and in some places, water rose well above their knees. Not surprisingly, the cost and the extreme difficulties of the enterprise exceeded the most pessimistic expectations, and after some years of fruitless efforts, work stopped. A group of "able and expert" masons and architects gathered to find a solution that would allow the continuation of the project. In 1570, they recommended that the new Cathedral be built on an adjoining site in which the ground was firmer due to the presence of ruins of the old Aztec structures. This preloaded site was more capable of withstanding the loads imposed by the new structure. The church axis was rotated and, instead of having an east-west orientation, as do most Christian temples, it is now aligned along the north-south direction. Most significantly, the commission also advised that rather than on continuous footings, the temple be founded on a masonry platform.

Construction of the Metropolitan Cathedral at the new site began in 1573 under the direction of Master Mason Claudio de Arciniega, who had faced the difficulties of construction in the marshy grounds of the city previously, most notably in the construction of the San Agustín convent and church.

Foundations

Following the recommendation of the learned group of experts, Arciniega built a masonry platform, really a foundation raft, with basalt, volcanic foam and lime mortar, over a surface of 140 × 70 m. Even though its average thickness is about 1.20 m, it is thicker towards the south, where softer ground is found, which suggests that the first builders added thickness at that particular zone to compensate the differential settlements that became apparent ever since the earliest stages of its construction. Before work on the actual raft began, the soil was reinforced by inserting some 22,500 wooden stakes or short piles of 20 to 25 cm in diameter and between 2.2 and 4 m in length. Archaeologists have revealed that the Aztecs, in their buildings, used this method previously to reinforce the marshy ground that prevailed

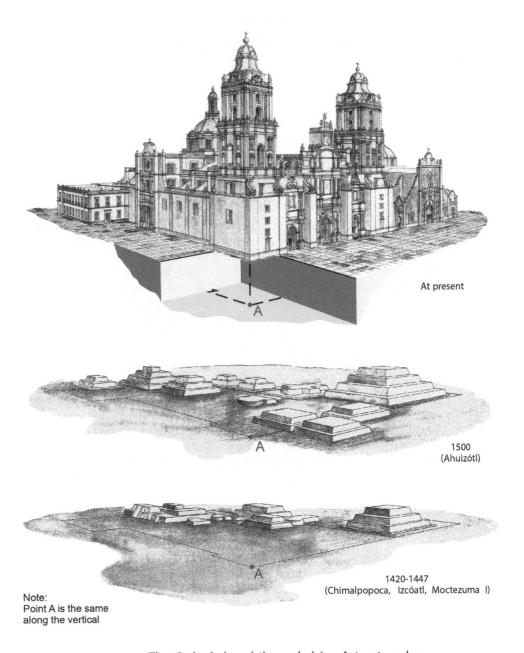

At present

1500
(Ahuizótl)

1420-1447
(Chimalpopoca, Izcóatl, Moctezuma I)

Note:
Point A is the same
along the vertical

The Cathedral and the underlying Aztec temples

Figure 2.1 The Metropolitan Cathedral and the previous pre-Hispanic structures.

in the city. On top of the platform a grid of masonry beams was built to receive the walls, pilasters and columns, as illustrated in Figure 2.2. The beams are 3.5 m in height, 2.5 m in width and as much as 127 m in length. The top part of the platform was at the same level as the Plaza Mayor (main square), and the grid of inverted beams reached a height of

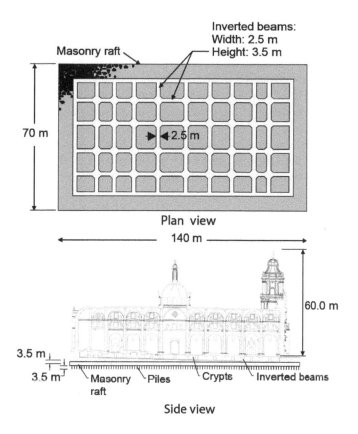

Plan view

Side view

Figure 2.2 The foundation of the Cathedral: masonry raft and a grid of beams built on top of about 22,500 short wooden piles.

3.5 m above this elevation, which is a clear indication that Master Builder Arciniega expected large-magnitude settlements to occur.

General structural characteristics

The body of the Cathedral occupies an area of 126.7×60.4 m and weighs 1,270 MN with an average contact pressure of about 166 kPa. It has five naves, the lateral ones closed by masonry walls and divided into chapels; the processional and central naves are limited by 16 stone columns. Horizontal thrusts from the vaults and the main dome are taken by the lateral walls and by perpendicular elements in the chapels that act like counterforts (see Figure 2.3). The central dome is supported by four columns, and the two bell towers are 60 m in height.

The vaults were erected next and were completed around 1667. The façade was finished in 1675, and Damián Ortiz de Castro finalized the towers in 1791. Manuel Tolsá profiled the dome and joined the complex with a balustrade and pinnacles as a characteristic architectural feature. Tolsá completed the work in 1813. This structural system has proven over the years to be well suited to carry intense seismic loads, to survive a five-year flooding during its construction and to accommodate exceedingly large differential settlements

Settlements in the Cathedral during its construction

The differential compressibility of the subsoil clay strata, due to the consolidation induced by the Aztec temples and structures pre-existing at the site, caused differential settlements from the beginning of the construction. These deformations produced in turn structural misalignment that was compensated during the construction stage by modifying the height of the columns and walls in order to level up the springing of the vaults. Architectural contrivances were also used to disguise the visual effect of the settlements, such as introducing variable heights in the cornices and using wedged quarried blocks at the two towers. The analysis of the geometrical details of the monument made it possible to demonstrate that during construction of the Cathedral, and prior to the completion of the vaults, column C-9 accumulated a maximum differential settlement of 85 cm with respect to the plinth of pilaster C-3 (Figure 2.3).

The Sagrario Church

The parish church known as the Sagrario was built directly on top of the pyramid of the Aztec sun god, Tonatiuh. The Sagrario was partially founded on the Cathedral's foundation platform, and its western wall is common to both structures. Its builder, Lorenzo Rodríguez,

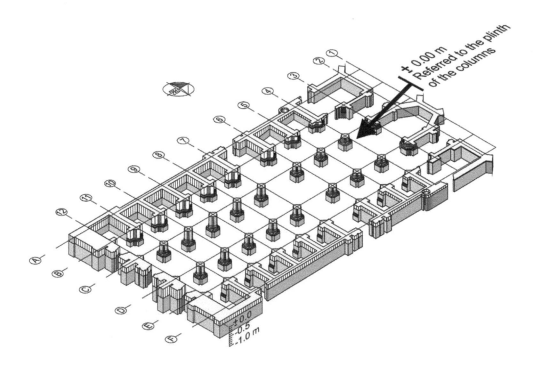

Enlargement of column shafts and walls during construction

Figure 2.3 Distribution of walls and columns in the Cathedral. The figure also shows differential settlements brought about during its construction.

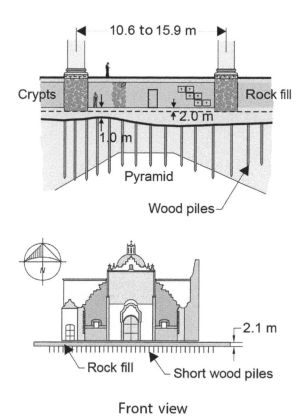

Front view

Figure 2.4 Schematic view of the foundation at the Sagrario Church, replicating the same foundation method as in the Cathedral.

also reinforced the soil with short woodpiles, as illustrated in Figure 2.4. On top of them a masonry platform was built but with lesser quality materials. The Sagrario has a Greek cross layout in plan with walls at the four corners that constitute the resistant part of the structure; its dome rests on four columns. The roof is formed by brick vaults supported by masonry walls, and the main central dome by four columns. It covers a square area of 47.7 m by side, weighs about 30,000 t (3,000 kN) and the average contact pressure transmitted is about 132 kPa. The construction of the Sagrario spanned from 1749 to 1768.

Other edifications

The old cathedral, dedicated in 1528, was located in the southern atrium, occupying most of its west side and demolished in 1625. The bishopric was erected in 1725 next to the western side of the apse. Part of this two-storey building, now a museum, was constructed directly on the Cathedral's foundation raft and its western portion founded on surficial footings. A small chapel was also built on the north-eastern side of the museum. Between 1795 and 1800, a seminary was built on the other side of the apse and was demolished in 1938. A plan of the Cathedral, the Sagrario church and the other edifices is also shown in Figure 2.5. The Sagrario church, the bishopric and the chapel interact with the Cathedral, contributing

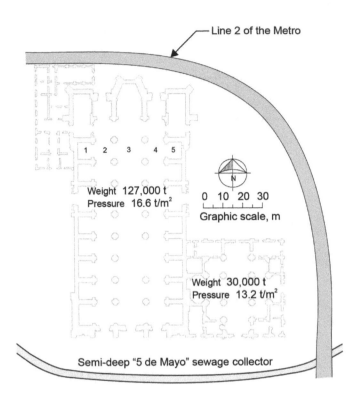

Line 2 of the Metro

1 2   3   4 5

Weight 127,000 t
Pressure 16.6 t/m²

0 10 20 30
Graphic scale, m

Weight 30,000 t
Pressure 13.2 t/m²

Semi-deep "5 de Mayo" sewage collector

Dimensions and weights of the Cathedral and the Sagrario

Figure 2.5 The Cathedral and the Sagrario Church and their proximity to two underground structures: the Mexico City subway and a sewage pipe near the south atrium.

to the complexity in the analysis of its movements. The old cathedral and the seminary also modified the stress history and the mechanical characteristics of the clay strata beneath them.

In 1968 a sewage collector was built at a depth of 16 m along the southern façade of the Cathedral and of the adjoining Sagrario. From piezometric measurements it has been inferred that this tunnel is permeable and that it drains out the water from the subsoil, particularly in the south-eastern zone of the Sagrario. Construction of Line 2 of the subway system (Metro) also started in 1968, and its cut-and-cover tunnel also acts as a drain at the north and east sides of both churches (see Figure 2.5).

Chapter 3

Foundation modifications

First intervention at the Cathedral

In 1929, the Technical Committee for the Conservation of the Cathedral appointed architects Manuel Ortiz Monasterio and Manuel Cortina García to prepare a project for reinforcing the beams in the foundation raft because they had undergone serious structural damages by differential settlements. The first corrective action undertaken was the demolition in 1938 of the Seminary building, with the purpose of unloading the eastern zone. Work at the foundations consisted of emptying the cells of the beam grid supporting the Cathedral, after which the average contact pressure decreased from 143 kPa to 108 kPa, i.e. a 25% reduction that was practically lost afterwards as a result of the weight imposed by the crypts that were subsequently built. In spite of the fact that the reduction of pressure generated a certain expansion of the underlying clay strata, its magnitude was soon counteracted and overcome by regional subsidence. With present geotechnical knowledge, it would have been concluded that such an action was not going to be efficient.

The project also envisaged the reinforcement of the masonry beam grid with the use of a vertical concrete and steel plate and the installation of structural steel beams that were supported by a concrete slab with an approximated thickness of 50 cm (as shown in Figure 3.1) that was only built at the east and west sides of the transept. The capacity of this system to compensate differential settlements turned out to be very limited because the section of the inverted beams was reduced to allow access to the crypts therefore weakening the inverted beams. Furthermore, the wood floor at the parish level was replaced by a reinforced concrete slab with a construction joint left along the column axis of the western side.

First intervention at the Sagrario Church

An attempt to underpin the Sagrario Church with 25 cm diameter woodpiles took place in the forties. The parish floor was reinforced with a concrete slab supported by a grid of steel beams. Later, between 1960 and 1964, another underpinning system was tried using concrete piles driven in 1 m lengths. Many of the top parts of such piles can be observed in the cells under the Sagrario; it is evident that a large amount of them could not be driven.

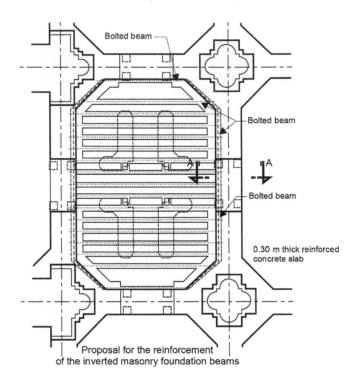

Proposal for the reinforcement
of the inverted masonry foundation beams

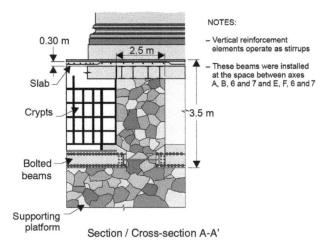

Section / Cross-section A-A'

First modification (Arq. M. Ortiz Monasterio)

Figure 3.1 First intervention in the foundation in 1929: reinforcement of the masonry beam grid with the use of a vertical concrete and steel plates and structural steel beams.

Second intervention

In 1972, the Mexican Government commissioned a geotechnical and structural study from which Manuel González Flores concluded that an underpinning based on his proprietary control piles could solve the problem. As seen in Figure 3.2, the piles are connected to the foundation slab through steel frames that control and regulate the magnitude of the loads applied at their heads; when the slab settles, pile heads tend to emerge.

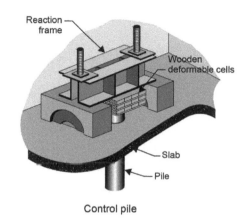

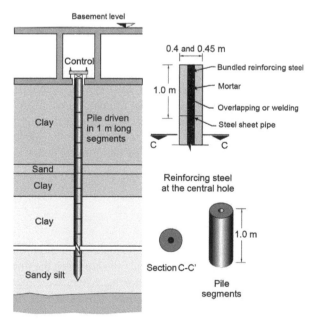

Figure 3.2 Schematic representation of a control pile and its components. The behaviour of the wooden deformable cells is very nearly elasto-plastic. Thus, the magnitude of the axial load transmitted to the pile can be regulated.

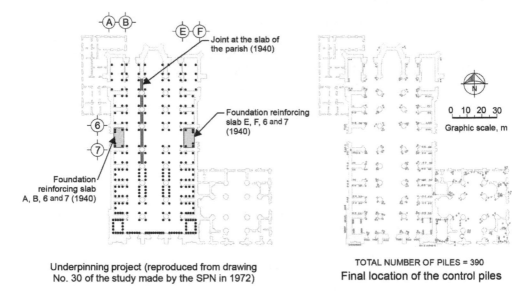

Underpinning project (reproduced from drawing
No. 30 of the study made by the SPN in 1972)

TOTAL NUMBER OF PILES = 390
Final location of the control piles

Figure 3.3 Initial project for underpinning the Cathedral with control piles (left-hand side). Final distribution of the control piles that were actually installed (right-hand side).

Underpinning proposal for a second intervention

The study performed by the SPN (an agency of the Mexican government) recommended the installation inside of the church of 280 piles driven to the first hard layer and placed mainly in the southern part, as depicted in Figure 3.3. It was also concluded that the control piles would reduce by 25% loads carried by the original foundation, would allow the adjustment of the settlement rate of the buildings with respect to the surrounding ground, and would correct differential settlements within the structures themselves. The difficulties encountered during driving of the piles led to a modification of the project, which consisted of driving the piles wherever possible and increasing their number to 390 at the Cathedral and to 129 at the Sagrario.

Classification of the control piles

The piles were classified on the basis of the lengths reported in the project logbook. A first group includes actual point-bearing piles. Those that failed to reach the first hard layer work as friction piles. Others longer than the depth to the first hard layer were regarded as inclined or otherwise as broken piles; in both cases they fail to have a reliable behaviour as point-bearing elements, since they are likely to work mainly as friction piles. Finally, long piles and friction piles that were installed outside the building are considered; they may have crossed the first hard layer due to the drilling-before-driving method that was used in the atrium zone. It was concluded in the study made in 1989 that only 27% of the piles are properly supported by the first hard layer, and at the Sagrario only 11% of the piles fulfil such condition (see Figure 3.4).

Assuming that only 103 of the piles installed at the Cathedral and 14 of those in the Sagrario are reliable, since they correspond to point-bearing piles, and considering that

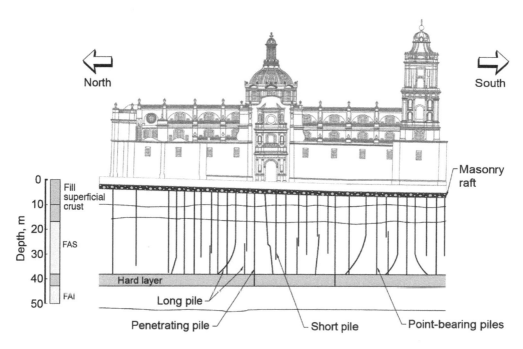

Figure 3.4 Classification of the control piles on the basis of the lengths reported in the project logbook.

their individual bearing capacity is 100 t, it can be concluded that a total bearing capacity of only 11,700 t is available. When compared with the approximate weight of both temples, 156,000 t, that capacity is only 7.5% of the required amount, obviously insufficient to modify the behaviour of the foundation of each of the churches. If, in addition, it is considered that the maintenance of these piles implies the task of lowering the supporting frames and the eventual shearing of the protruding pile caps, the contribution of the piles as controlling elements of the differential settlement in the long term becomes negligible.

Chapter 4

Characteristics of the subsoil

The soil that underlies the Cathedral and the Sagrario church has undergone a complex stress history as a result of stress changes imposed on it by a succession of pre-Hispanic and colonial buildings, by modifications to their foundations over the last five decades and by effective stress changes brought about as a consequence of pumping water from deep aquifers for nearly 200 years. Stress path-dependent parameters – strength and, more significantly, compressibility – vary considerably, both horizontally and vertically. Variations in these properties and the continued change in piezometric pressures brought about by pumping from the aquifer that underlies the soft clays are presently the main contributing factors for the generation of additional differential settlements.

Results of geological and geotechnical studies performed previously at the site were a most useful supporting material at the onset of the project. The information gathered during that initial stage allowed the authors to confirm that the original islet was only a small promontory that had a spring known to the Aztecs as Toxpálatl that existed under the west atrium of the Cathedral. Fresh water sprang from it, which explains why the Aztecs favoured that site for their initial urban settlement. By contrast, the water in the lake that surrounded that islet was a brine not suitable for human consumption.

Analyses of data gathered then were complemented with an additional geotechnical exploratory program that was carried out to define the details in the stratigraphy of the subsoil underlying the Cathedral and the Sagrario and to determine subsoil properties, in particular the compressibility of the materials. During the initial stage of studies in 1989, 21 cone penetration tests (CPT tests) were advanced as well as two borings with continuous undisturbed sampling. In the course of the construction of 32 shafts in 1993, 27 additional CPT tests were made, and two others in 1995.

In a CPT test a conical tip is driven into the ground at a constant penetration rate. An electronic cell is fitted above the tip to measure soil penetration resistance. This reaction is a function of two factors: a) the soil compressibility and b) the shearing strength of the soil itself. Tip penetration resistance measured with the electric cone is correlated to both factors.

Typical stratigraphic profile

Three CPT borings performed in front of the Cathedral and of the Sagrario made it possible to prepare the stratigraphic profile depicted in Figure 4.1, which shows that at the boundary between both churches the subsoil is stronger because it corresponds to the zone that has received the heaviest load transmitted by the Aztec temples, by an archaeological fill and by the two heavy colonial structures. In contrast, on both ends of the profile penetration

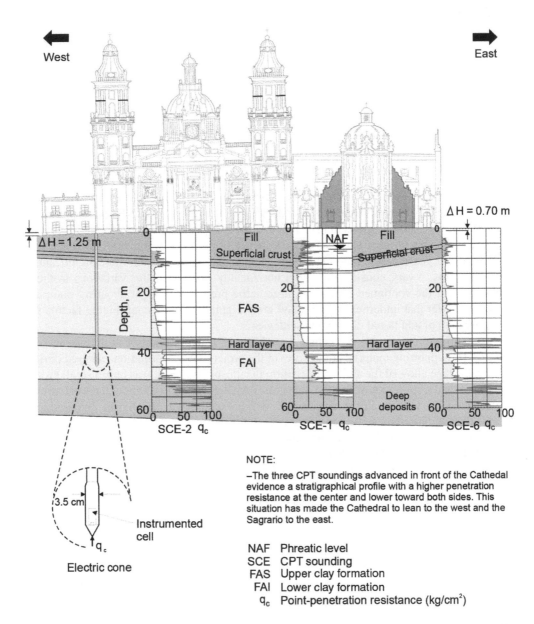

Figure 4.1 CPT borings performed in front of the Cathedral and of the Sagrario, interpreted to infer the stratigraphy at the site.

resistance is reduced almost by a half. This condition induced the tilting of the southern part of the Cathedral towards the west, whereas the Sagrario is inclined to the east. The same figure also shows the thickness and depth of the most relevant strata encountered in the soil sequence at the site.

Subsoil deformations

With the information derived from the CPT tests it was possible to define the depth of the contact between the natural shallow crust and the soft clays, a surface that was originally flat; however, as a result of the consolidation induced by the Aztec pyramids, it underwent depressions as deep as 10 m, as shown in Figure 4.2. For this reason, before the colonial structures were built, the site was levelled with artificial fills to shape a new initial plane. Laboratory tests (one-dimensional consolidation tests) demonstrated that the loads applied by the former pre-Hispanic constructions were released at some parts, although in other areas they were subsequently increased by the weight of the Cathedral and of the Sagrario. This complex load history brought about the heterogeneity in the conditions and properties of the subsoil that was detected in the field and laboratory tests, as illustrated schematically in Figure 4.3.

Measurements of underground water pressure in 1990

To complement the knowledge of the subsoil conditions at the site, pore-water pressures at different depths were measured in a piezometric station EP-1 installed at the southern atrium of the Cathedral. As seen in Figure 4.4, pore pressure was nearly hydrostatic between 0 and 20 m in depth; beyond this last depth a pressure loss of about 180 kPa was recorded at the first hard layer, found at a depth of 38 m, and of 200 kPa at the deep deposits, 53 m deep.

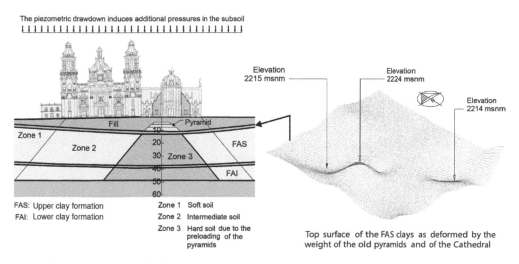

Figure 4.2 At right: effects of the application of long term loads on the compressibility of clays that underlie the Cathedral and the Sagrario Church. The graph on the right shows the configuration of the upper part of the clayey soils with the desiccated crust. The imprint left by the heavy pre-Hispanic structures is evident.

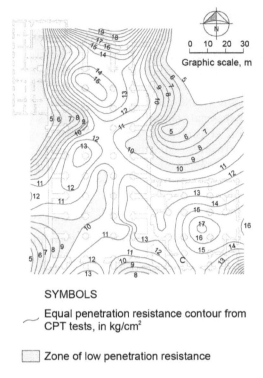

SYMBOLS

⌒ Equal penetration resistance contour from
CPT tests, in kg/cm²

▢ Zone of low penetration resistance

Figure 4.3 Heterogeneity in the distribution of estimated compressibilities of the subsoil identified from the results of field and laboratory tests.

PIEZOMETRIC CONDITIONS

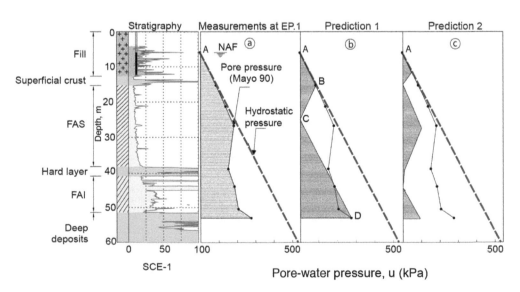

Figure 4.4 Pore pressure distribution of pore pressures measured in 1990 and two hypothetical distributions of pore pressure to estimate the evolution of settlements in the future.

Estimates of water pressure trends in the future

Pore pressure distribution recorded at piezometric station EP-1 can be expected to slowly decrease in the future, and that pore water may eventually define a hung aquifer formed by the infiltration of rainwater and by seepage from potable water and sewage mains. With these hypotheses, two predictions of the piezometric variation can be established:

- **Prediction 1**. It is feasible to imagine a suspended body of "trapped water" located between 6 and 25 m in depth as well as a hydrostatic distribution found underlying the former. This assumption implies a decrease of the hydraulic pressure down to a value of 180 kPa at the upper clay formation.
- **Prediction 2**. It could be also assumed that two suspended water levels will be formed, one of them between 6 and 13 m in depth and the other from 16 to 38 m. This implies pore pressure drops at such depths of 80 and 180 kPa, respectively. Furthermore, beyond a depth of 45 m, it may be also be expected to reach a hydrostatic distribution (Figure 4.4).

Chapter 5

Regional subsidence

Regional subsidence seriously affects urban infrastructure in Mexico City. Subsidence and subsidence rates vary depending on geographical location, site-specific geotechnical conditions, structural types and characteristics, and the mutual interactions between buildings or infrastructure elements. Hence, settlements and settlement rates are not uniform and, regarding the built heritage, non-uniform subsidence induces differential settlements that, in turn, gradually induce structural damage and also increase vulnerability to other phenomena, most notably to earthquakes.

Regional subsidence is brought about by the exploitation of the deep-seated aquifers that underlie the lacustrine clays throughout the former lakebed. Pumping of water induces a reduction of water pressure within the aquifer, which in general terms has two major characteristics: a) it is constituted by very pervious materials such as sand, sandy silt or gravel; and b) it is confined by low-permeability clays. As water pressure in the aquifer reduces, there is also a gradual decrease of the pressure in the water filling the voids of the clays. Depending on the thickness and on the permeability of the clay, a sudden change in water pressure in the aquifer produces deferred changes in the pore-water pressure of the low-permeability materials that may last even decades before a new state of equilibrium can be reached. Together with this, water will flow downwards very slowly from the clay into the aquifer.

If the clays are saturated, as it is very approximately the case of the soft clays in Mexico City, the volume of water ousted is proportional to the subsidence observed at the ground surface. Pressure changes undergone by pore-water pressures in the clay increase the stresses acting effectively on the solid phase of the soil. The compression of the latter follows, and it is because of this that the pumping process is equivalent to an effective surcharge of the soil, in response to the reduction of pore pressures. Regional subsidence in the city has two components that develop simultaneously, as in most clayey soils: a) primary consolidation through which interstitial water is expelled from the soil voids and predominates initially and b) secondary consolidation, a deformation process that persists for several decades.

When the effects of the regional subsidence became more evident, municipal engineers in Mexico City realized that in order to assess the magnitude of the phenomenon, it was necessary to have reliable topographical references.

Atzacoalco benchmark: basic reference for all topographic levelings in Mexico City. Neglect can be observed, and even the risk of destruction.

Topographic references

Figure 5.1 A view of the Atzacoalco benchmark in 1990.

Benchmarks

One of these references is a benchmark installed on rock outcrop outside the former lakebed and, consequently, not affected by regional subsidence. It is the so-called Atzacoalco bench-mark, which was established around 1860 and has traditionally been taken as the base reference for regional topographical works in the Basin of Mexico. In 1959 it was relocated nearby; thereafter, every major civil engineering project in the city uses it as the basic reference point to establish elevations in the city (Figure 5.1).

Recorded settlements

Javier Cavallari, an Italian architect who arrived in Mexico in the middle of the 19th century to join the staff of the School of Architecture in Mexico City, performed the first levelling between the Cathedral and the Atzacoalco Church in 1860. We have no information about the second levelling, but we do know that the third one was carried out by Roberto Gayol in 1892, taking as a reference the lower tangent of the Aztec Calendar (TICA) that used to be attached to the Cathedral's west tower. At that point in time, engineers and architects of the city considered that as a fixed point. But this is not so, since it is also affected by regional subsidence, as illustrated by analyzing the development of settlements at the TICA historical reference. This reference settled more than 8 m during the 20th century, and approximately 2.6 m over the last 35 years, as seen in Figure 5.2. The graph highlights the most significant values of yearly settlement rates. Between 1965 and 1990, subsidence varied almost linearly with time, approximately at a rate of 7.1 cm/ year. In the levelling made in 1991, this rate had decreased to only 1.6 cm/year, and subsequently, because of the effects of underexcavation at the Cathedral, it increased to 10 cm/ year from 1991 to 2000.

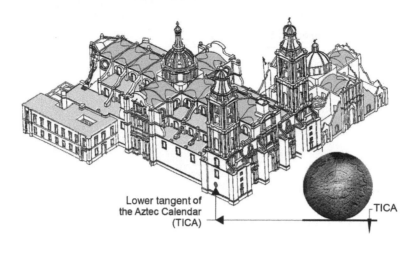

Lower tangent of
the Aztec Calendar
(TICA)

TICA

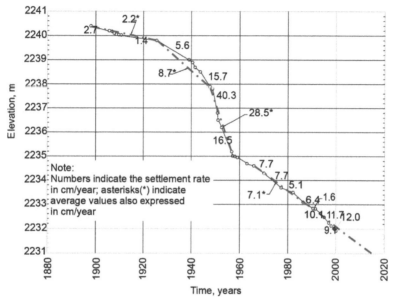

Regional subsidence of Tica reference at the Cathedral

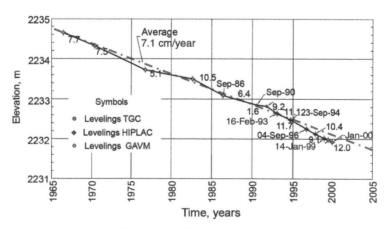

Settlement of the south-east corner

Figure 5.2 Evolution of settlement of the lower tangent of the Aztec Calendar, at the base of the west tower.

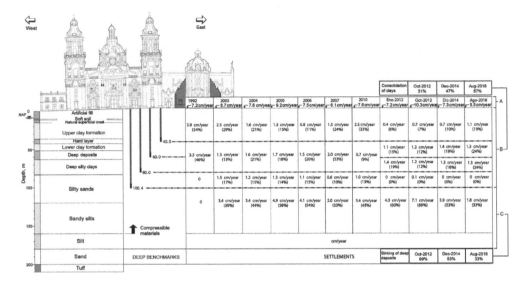

Figure 5.3 Contribution of the compression of the main soil strata to the total surface settlements measured over time at the Cathedral's west atrium.

Benchmarks to measure settlement distribution

As part of the underexcavation project for the Cathedral and the Sagrario church, deep benchmarks were installed to determine how settlements within the subsoil are distributed at the Cathedral and at the Sagrario, and also to measure the total regional subsidence. They are constructed with concentric twin pipes; the internal one acts as a reference mark, and therefore it is continuous and rests at the selected depth. The external pipe is deformable, and hence, it absorbs axially the vertical deformations suffered by the soil between the surface and the depth of the benchmark. The inner tube remains free, i.e. it is not affected by buckling, as are conventional benchmarks made with rigid outer pipes. Friction forces of the soil acting against the inner pipe are in fact eliminated.

Settlements measured at the deep benchmarks installed at the western atrium of the Cathedral, as well as the contribution in percentage of the major compressible strata, are indicated in Figure 5.3. Data given in that figure spans from 1991 to 2007. In 1991, compression from the upper clay formation contributed 54% of the total overall settlement, the lower clay formation 15%, and the deep silty clays of the former third lake the remaining 31%. The relative contributions of the main strata have changed over the years, as well as the magnitude of total settlements. The contribution of the uppermost clays has gradually decreased. In 2006, it was only 11%, thus showing the beneficial effect of mortars grouted into it. The graph also shows that the deeper strata are now important contributors to total settlements, a previously unexpected situation of serious concern because of the effect it will have on the deep sewage system of the city, albeit their contribution reached a maximum in 2012 and has been decreasing ever since. It stood at 33% on August 2018.

Chapter 6

Geotechnical diagnosis

In 1989 the Mexican authorities realized that deterioration of the Cathedral was unacceptable and that an intervention was inevitable if the monument was to be preserved. Differential settlements that year were 2.42 m between a point near the apse and the western tower, a huge amount that accumulated in the course of 419 years, beginning in the early stages of construction. Differential settlements are determined by the sum of two factors:

1   Consolidation induced by the weight of the pre-existing Aztec temples and of the subsequent colonial structure. The levelling performed in 1907 shows differential settlements that are mostly due to this factor, even though regional subsidence had already been triggered in the city but was not significant. The plot in Figure 6.1 shows differential settlement contours with respect to point C-3 near the apse thereafter considered as an arbitrary zero reference. Maximum differential was 153 cm that year.
2   Regional subsidence of the city. This has been the most important factor for the development of differential settlements during the last 150 years. It induced a differential settlement of 87 cm between 1907 and 1989 in the west tower with respect to the plinth of pilaster C-3 at the apse. Total differential settlement reached 242 cm in 1989, as seen in Figure 6.2.

In order to record and assess the effect of the regional subsidence in the development of differential settlements, several precision topographic levelling surveys were carried out at the Cathedral and the Sagrario during the stage of preliminary studies. The levellings were performed at the plane of the plinth of the columns supporting the Cathedral, therefore allowing continuity in the levellings of this surface that have been carried out since 1907. The plan in Figure 6.3 shows yearly settlement rate contours derived from measurements made between 7 January and 22 September 1991, shortly before underexcavation began. From the figure it is possible to infer the geometric deformations suffered by both churches during that time. These deformations represent the settlement trends that would have been observed had the underexcavation project not been carried out. As an example, the western bell tower settled at a rate of 12 mm per year with respect to the central part of the nave; the southeastern corner of the Sagrario was settling 16 mm with respect to its central part, and the vertical deformation of the museum building was 26 mm, taking as a reference the main altar.

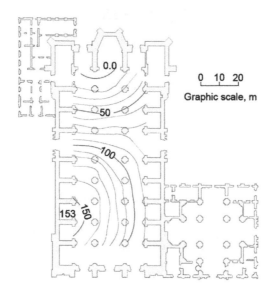

October 1907
Contours prepared with data from SPN

Figure 6.1 Results of the levelling performed in 1907, with differential settlements with respect to point C-3.

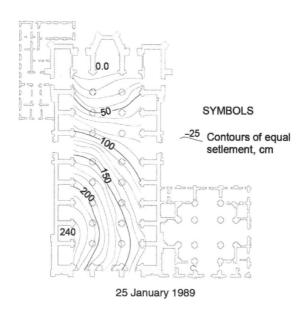

25 January 1989

Contours prepared with data from PICOSA
Evolution of past differential settlements, cm

Figure 6.2 Accumulated differential settlements with respect to point C-3, measured at the beginning of the project in 1989.

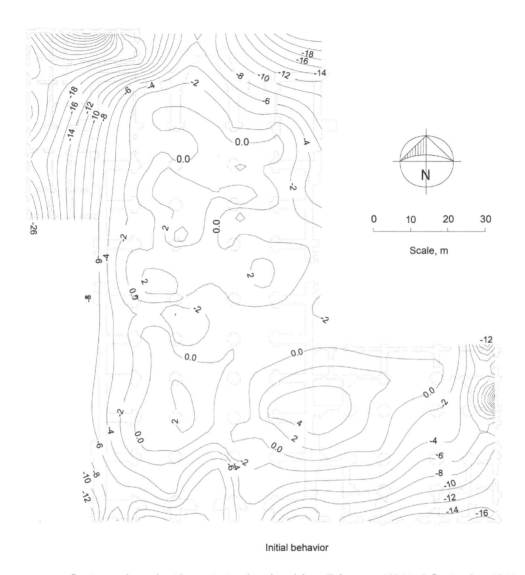

Initial behavior

Contours of equal settlement rates (mm/year) from 7 January 1991 to 2 September 1991

Figure 6.3 Initial differential settlement rates in mm/year as measured from January to September 1989.

Calculation of the initial settlements

Settlements induced by the Aztec pyramids at the zone where the Cathedral and the Sagrario were subsequently built were estimated after assuming the probable thickness that the soil strata had under both churches prior to the construction of the pre-Hispanic structures. The method applied to define the initial subsoil condition is similar to that used by professors Mazari, Marsal, and Alberro to reconstruct in 1984 the stress and strain history of the subsoil under the Great Aztec Temple (Mazari *et al*., 1985). Exceedingly large deformations ranging from 7 to 13 m were obtained for the conditions that were assumed to prevail in the 14th century and thereafter. Loads, areas subjected to stresses, thicknesses of the deformable materials and the compressibility parameters at that time were estimated from archaeological data and historical records. These values agree reasonably well with those deduced from the results of geotechnical exploratory borings performed at the site, mainly from CPT tests.

Prediction of future differential settlements

A forecast of long-term settlements was carried out using traditional procedures and methods used in soil mechanics. This prediction was made assuming that the churches would be left as they were in 1989, i.e. without intervention. As a result, we obtained a plausible picture of the consequences that could be faced had they not been rehabilitated. Future settlements of the Cathedral and of the Sagrario would have been governed by the evolution of the pore-water pressures in the clay deposits. Two hypotheses were assumed for the future hydraulic conditions likely to prevail in the subsoil, as mentioned earlier and illustrated in Figure 4.4. Prediction 2 leads to more unfavourable estimates of the future differential settlements because the differential cumulative movement at the western bell tower could have reached 3.2 m. On the other hand, in the case of the Sagrario, the average differential settlement between the central zone and the corners could be of 1.2 m. The configurations of the estimated future settlements are presented in Figures 6.4 and 6.5 for predictions using the two hypotheses for the evolution of future settlements. The maximum values would have occurred at the western tower, whereas the smallest movements would develop at the central part of the Sagrario. Estimations of the future evolution of settlements in the central part of the city, in and around the Cathedral, have also been carried out using an elasto-viscous-plastic constitutive law for the clayey soils, as seen in Figure 6.6, which shows settlement estimated for a 40-year period, from 1990 to 2030 (Ovando-Shelley *et al.*, 2007).

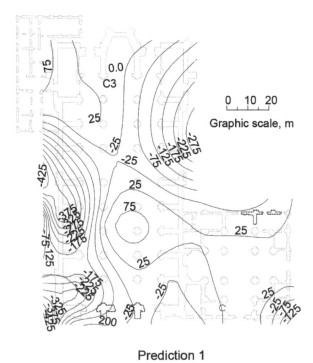

Prediction 1

Figure 6.4 Future settlements estimated with the pore pressure distribution of Prediction 1 of Figure 4.4.

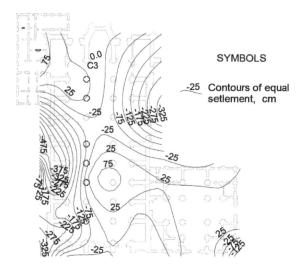

Prediction 2

Estimate of future settlements,
in cm, induced by regional subsidence

Figure 6.5 Estimated settlements calculated from Prediction 2 of Figure 4.4.

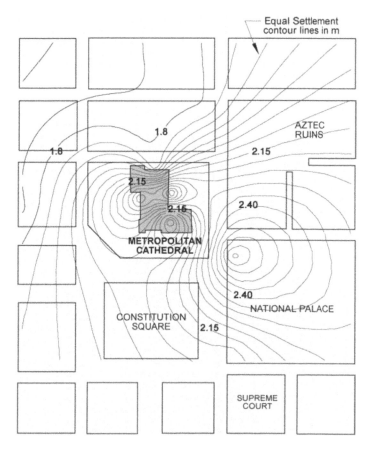

Figure 6.6 Future relative settlements in and around the Cathedral for the year 2030, calculated using
a viscous-plastic constitutive model.

It is evident that large magnitude differential settlements would induce inclinations and tilts as well as fissures and cracks induced by the appearance of tension stresses within the masonry of both the Cathedral and the Sagrario church. In general, structural damage would increase their vulnerability to earthquakes. A large magnitude earthquake such as the one that occurred in 1985 could generate a stress condition that could pose a serious risk to the stability of the churches, particularly that of the western tower, taking into account the distortions that the structures already had together with the future deformations induced by regional subsidence. When the next large earthquake hit the city in 2017, local intensity was lower at the site, and there were no major structural damages in these two monuments, although some ornaments fell off or underwent significant damage.

Chapter 7

Solutions analyzed

Solution alternatives

Several methods and criteria for actions to correct the differential settlements that existed prior to the project were proposed and analyzed. The characteristics and merits of these proposals were pondered in the short and long terms, taking into account their capabilities for mitigating or reducing the apparition of future differential settlements.

Negative skin friction piles

These are rigid inclusions inserted into the ground without being connected to the superstructure. This is a foundation solution that has been used in the city lately, but at the onset of the project in 1989 it had only been used in just a few cases. Conceptually, it was considered that these piles would reinforce the underlying soft clay, creating a block of reinforced soil by inserting a large number of point-bearing piles (approximately 1,500), driven to the first hard layer and capable of indirectly supporting the total weight of the surrounding ground and of the Cathedral itself. This solution would induce the emergence of the religious structure with respect to the surrounding ground level, as it has occurred in other structures in the city, the Monument to the Revolution or the Monument to Independence, for example.

Drilled shaft foundations supported by the deep deposits or by the first hard layer

With this solution, the settlement of the structure would be stopped by isolating it from the supporting soil through the construction of drilled shafts (about 240) excavated to the deep deposits and connected to the original foundation by means of mechanical devices that would allow the adjustment of the structure, as necessary, in order to correct existing tilts and to prevent inclinations from increasing in the future. The piers would be capable of supporting all the weight of the Cathedral and the negative skin friction generated when the soil settles due to the regional subsidence. With this technique, it would be possible to lower the whole structure under controlled conditions, therefore isolating it from the differential regional subsidence. It would imply the reinforcement of the present foundation to transmit the loads to the piers and the superstructure of both churches. Its construction is quite complex due to the presence of the control piles already installed and to the need of employing cumbersome construction equipment that would have to be operated from the parish floor, therefore obstructing the operation of the chorus and of the lateral chapels of the Cathedral.

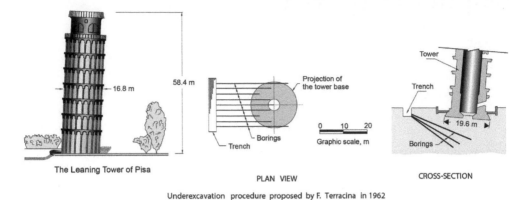

The Leaning Tower of Pisa

PLAN VIEW   CROSS-SECTION

Underexcavation procedure proposed by F. Terracina in 1962

Figure 7.1 The underexcavation project put forth for the Tower of Pisa by the Italian engineer Fernando Terracina in 1962.

Underexcavation

In 1962, the Italian engineer Fernando Terracina proposed a technique now called underexcavation to straighten the leaning Tower of Pisa (Terracina, 1962). The basis for his proposal is illustrated in Figure 7.1. His objective could not be fulfilled, but some years later his proposal was applied to several buildings in Mexico City. At the San Antonio Abad Church, it was used experimentally to demonstrate its feasibility for the Metropolitan Cathedral (Ovando-Shelley *et al.*, 1994). Basically, the purpose of this procedure is to accelerate the settlement rate of the "hard (more consolidated) zones" to prevent them from emerging with respect to the "soft (less consolidated) zones". This was achieved by extracting, under controlled conditions, through horizontal or inclined borings, the soil over which the foundation rests.

Pore-water recharge

Since regional subsidence in Mexico City is ultimately caused by the depletion of pore water pressure inside the compressible clay masses, a possibility was analyzed to decrease the magnitude of the future settlements through artificially recharging water into the pervious subsoil strata therefore stopping the consolidation of the soft clay strata. An analysis was made of the brief experience gained when this technique was used at two of the corners of the National Palace (Moreno-Pecero, 1981). For its implementation at the Cathedral, 46 absorption wells would be required, spaced every 10 m, as well as the construction of an impervious cut-off wall made with a plastic mortar trench that would confine the Cathedral and the Sagrario (see Figures 7.2 and 7.3). Within this enclosure, water would be injected by means of infiltration wells. The aquifer recharge would at the most control 69% of the settlements because the treatment would be applied to the two topmost clay formations, and it would become effective provided the water injections would last permanently; otherwise, the settlements thus prevented would inevitably occur again. In spite of its shortcomings, the aquifer recharge and the peripheral cut-off wall should be considered as a complementary eventual measure.

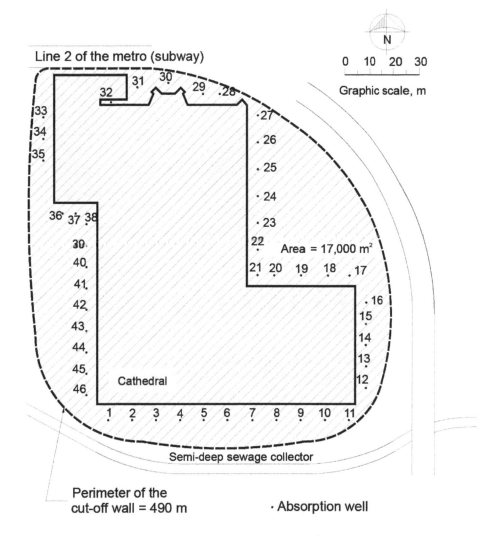

Line 2 of the metro (subway)

0 10 20 30

Graphic scale, m

Area = 17,000 m²

Cathedral

Perimeter of the
cut-off wall = 490 m   · Absorption well

Permeability = 10⁻⁷ cm/seg

Location of the absorption wells and of the cut-off wall

Semi-deep sewage collector

Figure 7.2 Conceptual project for isolating the Cathedral from the effects of regional consolidation by means of a peripheral cut-off wall.

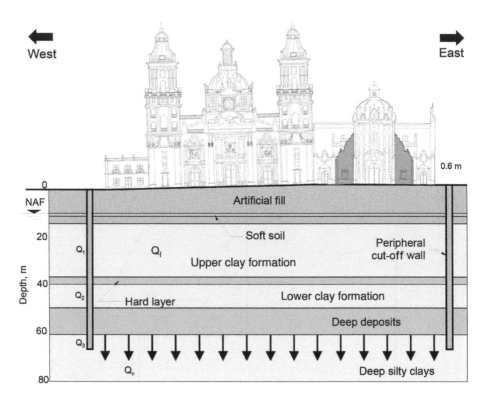

Flow rates lost at the bottom of the area confined
by the cut-off wall ($Q_t = 147$ m³/day)

Figure 7.3 Front view of the peripheral cut-off wall along the south atrium.

Other solutions

Other deep underpinning techniques were analyzed, such as the "pali radice" (Lizzi, 1983), constituted by inclined small-diameter micro-piles that, when intertwined, create stiff blocks to transmit the loads to the deeper strata. Such a layout would also solve the problem if applied under the whole area covered by the Cathedral and the Sagrario. This solution also required reinforcement of the existing foundation to enable it to support the micro-piles. It also implied accepting the inevitable apparent emersion of the churches.

All remedial works for deep underpinning projects based on piles, drilled shaft foundations or "pali radice" have as an added shortcoming that their complex construction procedure would imply closing of the Cathedral during many months, an unwanted situation for the religious authorities.

Chapter 8

Underexcavation at the Cathedral and at the Sagrario

This treatment was implemented to reduce differential elevations and tilting. It began in June 1993 and went on for slightly more than five years until August 1998. The technique involves lowering of the high parts with respect to the low points through the slow and controlled extraction of the soil that underlies the foundation. Three specific tasks were needed to underexcavate the Cathedral and the Sagrario Church:

1 The construction of access shafts to reach the upper clay formation, located between 14 and 21 m in depth.
2 The punctual drawdown of the phreatic level.
3 The underexcavation or controlled extraction of small portions of soil until a pre-established volume had been removed.

The two first operations are preliminary, whereas the third one constitutes the corrective geotechnical procedure itself, which was achieved by drilling 10-cm diameter radial borings advanced from the bottom of the access shafts.

Shafts for underexcavation

There were 32 concrete-lined cylindrical vertical wells to gain access to the level where the soft clays were detected. Their number and location, illustrated in Figure 8.1, were defined by applying approximate analytical methods to assess the effects of underexcavation.

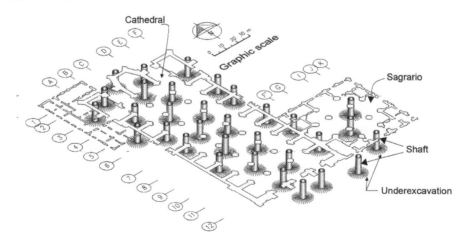

Figure 8.1 Distribution of shafts for underexcavation.

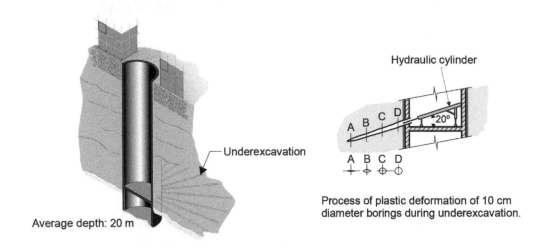

Shafts for the underexcavation process

Figure 8.2 Underexcavation process from the bottom of the shafts.

Underexcavation level

Soil extraction was carried out from the bottom of the shafts, in the soft clay at the contact with the upper clay formation, as shown in Figure 8.2. In each of the shafts a maximum of 50 radial borings penetrated into the soil in lengths ranging from 6 to 22 m. Figure 8.2 also shows an illustrative cross-section of one of them and a profile of the lengths penetrated with the underexcavation, as well as the gradual closure of those borings to induce the required settlements.

Pumping system

This system was applied during the excavation of the shafts in order to achieve the gradual drawdown of the phreatic level and to prevent bottom failure. Pumping was maintained throughout the soil extraction process. Inside each of the shafts, four-point wells attached to injection hoses were installed to extract water, which was conveyed to the sewage network. Pumping also induced corrective settlements.

Working equipment

A rotating steel table supported against the walls of the shaft was used to support pumps serving hydraulic cylinders close to the bottom of the shafts, with which excavating pipes were inserted and pushed into the clay. These cylinders are similar to thin walled soil sampling tubes. The photograph in Figure 8.3 shows a view of one of the shafts and a hydraulic cylinder for soil extraction is shown in Figure 8.4.

Figure 8.3 Photograph showing a soil extraction shaft. A device for soil extraction can be seen at the bottom.

Figure 8.4 Hydraulic cylinder for extracting soil.

Figure 8.5 Structural bracing installed as a safety measure.

Structural bracing

Underexcavation was executed with the aid of a complex preventive bracing system for the purpose of controlling any unexpected deformations and to prevent any structural damages (Figure 8.5). This system operated during the full process to adjust it to the gradual changes induced.

Underexcavation control

The weight and the moisture content of the material extracted were accurately and rigorously monitored throughout the process. In each of the pipe sections recovered with soil inside, either partly or fully filled with clay, the net weight and the water content of the material extracted were carefully recorded. In addition, it was a routine activity to recover soil samples for laboratory testing to obtain a reliable estimate of the underexcavated volume as well as of its mechanical properties. The distribution of the volumes extracted is illustrated in Figure 8.6; the overall volume excavated between August 1993 and June 1998 was 4,219.5 m³, extracted in 1,186,398 excavating actions. This volume was selectively distributed to comply with the correction targets.

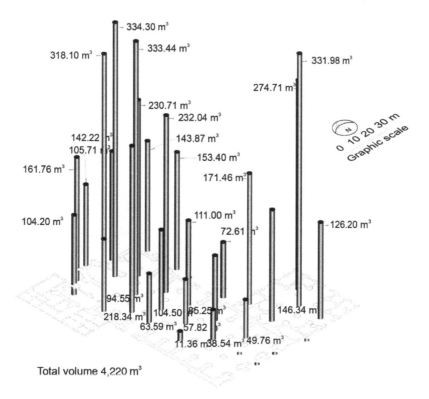

Figure 8.6 Volume of soil extracted from each of the 32 shafts. Total volume of extracted soil: 4,220 m³.

Correction targets

Figure 8.7 shows the targets to be achieved by applying the underexcavation method. One of them was proposed by Dr. Fernando López Carmona from the School of Architecture, and the other was put forth by Dr. Roberto Meli Piralla, from the Institute of Engineering, both from UNAM, the National Autonomous University of Mexico. In the first target the basic idea was to close the cracks and gaps that affected the Cathedral and the Sagrario Church. The Cathedral was to rotate towards the north-east and the Sagrario Church towards the north-west. The second proposal postulated that the Cathedral should undergo a rigid body rotation towards the north-east as well as the western portion of the Sagrario Church, whereas its east side would rotate towards the north.

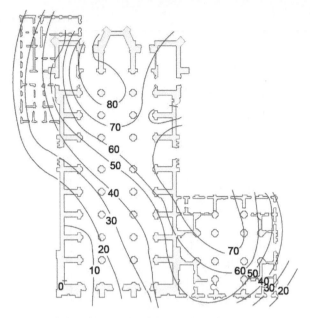

a) Dr. Fernando López Carmona

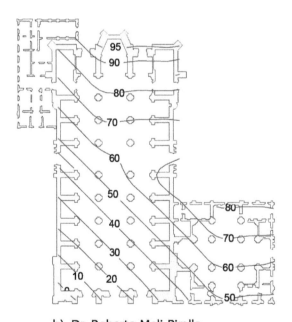

b) Dr. Roberto Meli Piralla

Symbols

20 ⌒ Contours of equal settlement, cm

Contours of proposed target settlements

Figure 8.7 Correction targets put forth by Professor López-Carmona (top figure) and Professor Roberto Meli (bottom figure).

Chapter 9

Geometrical correction achieved

By the end of June 1998, underexcavation basically eliminated differential settlements accumulated over the previous 65 years as a result of regional subsidence. When the treatment was completed, the maximum correction induced was 92 cm, between the apse and the southwestern corner. However, in September 1999 the maximum corrective settlement reduced to 88 cm and to 30 cm at the Sagrario, as shown in Figure 9.1. The reduction of corrective settlement from 92 to 88 cm was brought about, as discussed previously, by the fact that upon the end of underexcavation and the stoppage of the pumping operations, the effects of regional subsidence returned again, and as a result, part of the corrective settlements that had been achieved was lost. The net results were positive. For example, differential settlement between points C-3 and B-10 changed from 243 cm in 1989 to 156 cm by June 1998. The average angular correction between these two points was 23' 06" and of 25' 33" between points D-1 and A-12. Comparing Figures 8.7 and 9.1 shows that the targets set were fulfilled with underexcavation.

Geometry of the corrected settlements

The configuration of corrective settlements is represented with meshes as those depicted in the upper part of Figure 9.2. The upper mesh shows cumulative and corrective differential settlements as of August 2000. The lower part of the figure presents the same corrective settlements but referred to a horizontal plane, i.e. without considering historic differential settlement. The shape and distribution of the corrective settlements that reached a maximum of 88 cm is also shown.

Corrections at the towers

Both towers were monitored periodically to record changes in inclination starting in October 1993 using electronic plumbs with a sensitivity of 0.1 mm and 35 m long steel wire lines (precision = 1/35,000 or 5.9"). The correction achieved by April 2000 at the west and east towers consisted in rotations and displacements of 28.7 and 27.9 cm, respectively, along a northeast direction. At the beginning of the project the west tower was 109 cm out of plumb, leaning towards the south-west; underexcavation induced a rotation in the opposite direction which amounted to a correction of 26.3 % in the initial tilt. In the east tower the initial tilt was 113 cm, and underexcavation induced a corrective tilt of 24.6%.

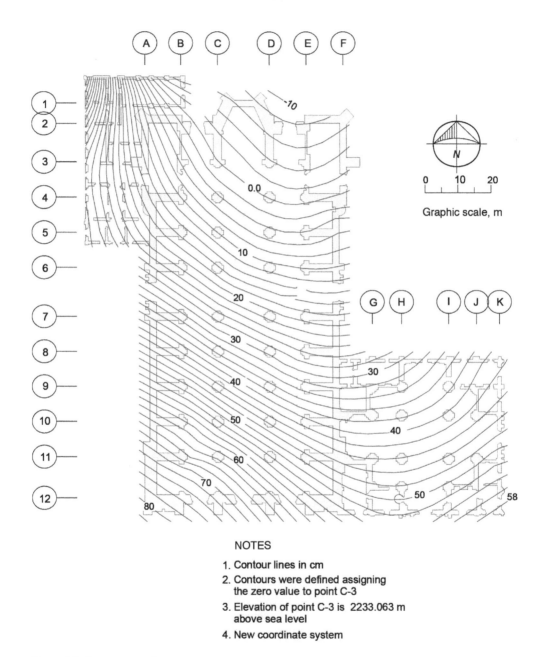

NOTES

1. Contour lines in cm
2. Contours were defined assigning the zero value to point C-3
3. Elevation of point C-3 is 2233.063 m above sea level
4. New coordinate system

Figure 9.1 Corrective differential settlements with respect to point C-3 induced by underexcavation (data taken from 25 October 1991 to 20 September 1999).

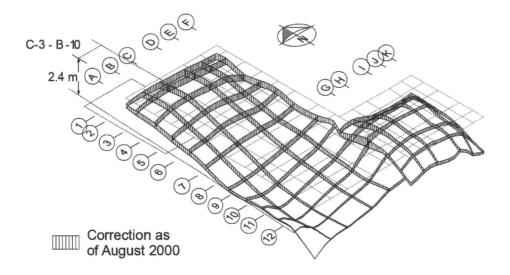

C-3 - B -10

2.4 m

Correction as
of August 2000

a) Isometric mesh of differential
   settlements and induced corrections

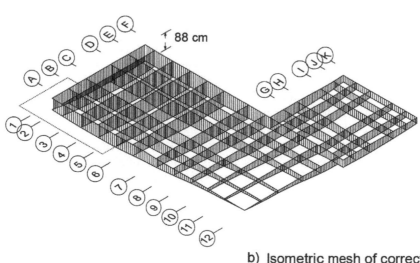

88 cm

b) Isometric mesh of corrections
   referred to a horizontal plane

Geometry of corrected settlements

Figure 9.2 Geometry of corrected settlements.

Reference plumb line

The impossibility of seeing with the naked eye the corrections that were in course prompted the installation of a plumb line at the Cathedral's central dome in order to make its movements apparent to visitors. The path followed by the projection of the central dome's geometrical centre in the course of time, beginning in 1573, is indicated by the trajectory of the plumb line, as shown in Figure 9.3. The initial path described a straight line in the south-west direction which changed rather sharply towards the west a few years after 1847, at a date that coincided with the opening of local wells for the extraction of water from underground aquifers. From October 1989 to October 1994, the dome leaned towards the north-east; subsequently, after adjusting the underexcavation program and up to June 1998, it rotated mostly to the north. As it can be viewed in the same figure, the total correction of the dome's tilt was 31 cm; this value is equivalent to a corrective angular rotation of 25' 18".

Structural damages

Underexcavation induced two types of movements:

1   The lowering of the central part of the Cathedral to recover the confinement provided to the vault by the walls.
2   Rigid body displacements towards the northeast.

The effects of the first type of movement at the beginning of the project could be verified because measurements indicated that the vault rose several centimetres. Subsequently, an adjustment was made to the sequence of underexcavation to achieve a rigid body movement.

Corrective settlements reduced crack widths and tilts in columns. However, new cracks developed, and others that already existed widened. In addition, plastering fell off at some points. Nonetheless, damages were considerably smaller than those expected at the beginning of the work. The condition of the structure was recorded and logged with plumb lines and conventional deformation gauges and also with an electronic continuous monitoring system. Analyses showed that the safety conditions of the churches were at no time at all in the course of the project in a situation of risk (Meli and Sánchez, 1995). The shoring, the confinement reinforcing at the columns (splints) and the turnbuckles that were installed at the roof acted as a protection to the structures against possible damage.

The most critical aspect in what refers to the safety of both churches concentrated on the columns, and it was therefore decided to grout them to achieve long-lasting improvements in their safety factors. Since the harmful effects of regional subsidence reappeared when underexcavation was stopped in June 1998, a small fraction of the corrections already achieved started to be gradually lost, but at mid-year 2000, the beneficial effects of mortar injections in the subsoil began to be evident.

Comparison with the underexcavation at the leaning Tower of Pisa

Construction of the Tower of Pisa began in 1173, almost exactly 500 years before the Mexico City Cathedral. The Tower is a hollow cylinder weighing 14,500 tons, slightly over 58 m in height and close to 13 m of external diameter. Its walls, with a thickness of 4.09 m, accommodate the spiral stairs to reach the campanile; they are made of masonry lined with marble.

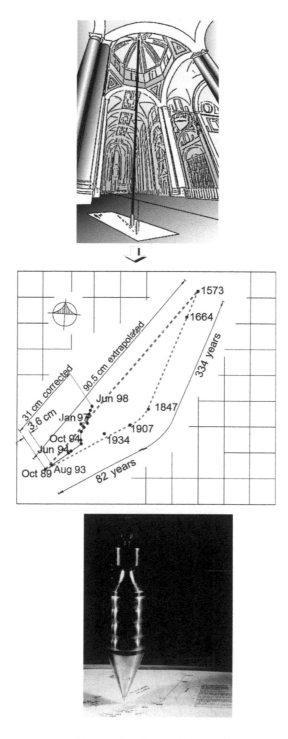

Reconstruction of the path
followed by the dome

Figure 9.3 Path followed by a plumb line installed in the main dome before and during underexcavation. Historical reconstruction of the path obtained from architectural and historical archive.

The historic pathology of the Tower of Pisa derives from the subsoil characteristics that, as in the case of the Cathedral, evidence non-homogeneous properties. This matter has been discussed amply by other distinguished researchers and engineers (Jamiolkowsky *et al.*, 1999; Burland *et al.*, 2002; Burland *et al.*, 2013) In 1990, inclination rate was 5.4 inches per year, and its total tilt was 5.5 degrees, equivalent to 9.7% of deviation from the vertical (Jamiolkowski *et al.*, 1993; Jamiolkowski, 2001).

The tower was also underexcavated, having as a precedent the underexcavation of the Mexican cathedral that had begun in 1993. An in situ underexcavation trial began in the Italian tower in 1999, six years after the Cathedral. This initial trial went on from February to June, during which 7 m^3 of soil were extracted. Full underexcavation took place between February 2000 and June 2001, during which the amount of soil extracted was 38 m^3. Techniques and mechanical gear used at the Tower were different from those used in Mexico. Indeed, these two cases show that the procedure can be adapted to meet the specific needs and limitations of individual projects.

At the Cathedral, the 88 cm of corrective settlements were equivalent to eliminating the vertical displacements accumulated during most of the 20th century for nearly 65 years. In the case of the Tower of Pisa, underexcavation reduced approximately 17 cm of the total cumulative differential settlement of 185 cm, i.e. it eliminated the vertical displacements that had accumulated in the previous 140 years. The volume of soil excavated at the Tower was about 30 m^3, whereas the volume extracted at the Cathedral amounted to 4,220 m^3.

The Metropolitan Cathedral would have been exposed to the detrimental effects of regional subsidence had no preventive measures been taken when underexcavation finished. Maximum differential settlement rate at the end of underexcavation might have reached about 1.7 cm/year, many times faster than the expected differential settlement rate for the Tower after it was underexcavated. In the specific case of the Cathedral's west tower, the expected settlement rate might have been about 20 times faster than the Pisan Tower, a situation of major concern since increasing large tilting induces increments of seismic vulnerability.

Chapter 10

Assessment of mortar injections

In looking at the Cathedral's long-term behaviour within the context of its continued exposure to regional subsidence, it was convenient to analyze other solutions to eliminate or at least mitigate its consequences. One of them was to accept the need for having periodic underexcavation interventions, a corrective measure that does not deal with the causes of the problem at the Cathedral, only with its effects. Accepting that regional subsidence will be present in the former lakebed in the city, at least in the foreseeable future, and keeping in mind that differential settlements are caused by a non-homogenous distribution of soil compressibility, it was decided to selectively harden the underlying clays to achieve a more uniform distribution of compressibility within the underlying clays. In other words, accepting that total settlements are unavoidable, selective hardening attempts to eliminate or at least mitigate differentials.

Background

The Palace of Fine Arts (Palacio de Bellas Artes), formerly the Teatro Nacional, is another iconic building in Mexico City; its construction began in 1905. Differential settlements appeared early on and were first noticed in 1906, when its foundation slab was being constructed. This led to a decision to harden the soft underlying clays by injecting mortar grouts, which were performed from 1910 to 1925, first with a cement grout and later on with a fluid mortar possibly made with lime and sand. The objective sought for with those injections was to radically arrest total settlements, although to no avail. The photographs in Figure 10.1 show the Palace of Fine Arts as it was being built in 1906, 1910 and at a later date.

Even though the grouts did not impede the appearance of total settlements, it should now be acknowledged that a great success was achieved because they became uniform, and damages to the structure of the theatre were avoided. It is worth mentioning that the same procedure was applied in 1881 at the Buenavista railway station. It is unfortunate that no information survives about that pioneering work carried out more than a century ago to reduce the settlements of a building in Mexico City.

Introducing grouts into the subsoil was questioned because many engineers doubted its efficacy. It was also misinterpreted because it was thought that the soil would be impregnated with the cement grout. During that time, regional subsidence in the city had not been acknowledged. The reader should also bear in mind that soft soil engineering at that time in Mexico City was essentially based on previous experiences, and that soil mechanics as an independent and recognized discipline of civil engineering hadn't even been born.

Photograph taken in December 1906, when the
differential settlement started causing alarm.

Photograph taken in August 1910, when the
subsoil grouting started.

A recent photograph.

Figure 10.1 Photographs showing the Palace of Fine Arts as it was being built. Top photo taken in 1906,
middle picture in 1910 and a recent photo at the bottom.

When grouting stopped after 1925, the case lost momentum since political unrest of those years diminished interest in the topic, and finally the theatre was left unfinished until the late thirties. The technical information was filed, and the unfair remark that "grouting failed to serve its purpose" was the only judgment about this experience that remained. This almost forgotten experience is a remarkable precedent of the method developed early in the 21st century to modify the compressibility of the subsoil under the Metropolitan Cathedral.

Theoretical and experimental studies

Theoretical and experimental research began in 1997 to study the effect of mortar grouts injected into soft clays in order to selectively reduce their compressibility. Experience gained from mortar injections at the Palace of Fine Arts was considered, and an investigation was made into the expected volume and characteristics of the grout. Field tests were carried out at the former Texcoco Lake bed, as well as complementary laboratory tests, theoretical studies on the hydraulic fracturing phenomenon and numerical simulations of the effect of introducing fluid mortars in the subsoil. Field tests were made sequentially to profit from the knowledge that was progressively acquired; at the beginning, the test results were poor, but they improved with time until they became reliable.

These investigations showed that when a fluid is injected under pressure inside a very soft mass of clay, such as that found in Mexico City, fissures and cracks are induced along planes whose orientation depends on the state of stress existing in the ground. Injected mortar penetrates through those cracks, forming vertical or nearly vertical sheets that configure a structure which reduces the compressibility of the soil mass. This results from the fact that the major principal stress in normally consolidated or slightly overconsolidated soils is vertical, or nearly so. This mechanism is familiar to geotechnical engineers as hydraulic fracturing, and it has been studied and described in investigations related to the behaviour of earth embankments and tunnels excavated in soft soils.

Field tests results

To verify the formation, length and thickness of the cement/lime mortar sheets, a 4.2 m diameter shaft was excavated to a depth corresponding to the treatment layer (Figure 10.2). Its walls were stabilized with shotcrete reinforced with electrically welded wire mesh. Point wells were also installed to allow the excavation of the shaft to a depth of 6.15 m. By advancing the excavation in 50 cm stages, it was possible to observe in full detail the grout sheets that covered lengths from 1 to 3 m.

Figure 10.3 shows a schematic cross-section of the geometry of the mortar sheets that were formed during this field test. Finally, from these tests it was possible to improve the techniques for determining accurately the amount of mortar grout necessary to generate the structure of a cluster of reinforcing sheets. Pictures of the injection of the core and of the mortar sheets already formed are also given in Figure 10.3. The soundness of the mortar core could be verified; it reached an average diameter of 29 cm, thus implying a radial expansion equivalent to 26% of the core drill diameter.

Formation of a mortar sheet

0 m
-3 m

-10 m
-13 m

-33 m

Rock fill

Artificial fill

Natural crust

Sheet

Upper clay
formation

Core

Hard layer

Lower clay
formation

Assembly of sheets

Structure of the injected mortar

Figure 10.2 Grouting experiment carried out at a site in the former Texcoco Lake north of Mexico City. Soils at the site have the same compressibility as those in the city centre.

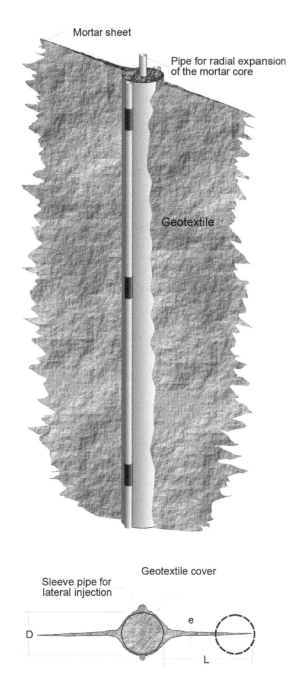

Core with lateral sheets of injected mortar

Figure 10.3 Schematic view of an injection core and of a mortar sheet.

Chapter 11

Hardening the subsoil

The results of the field grout tests in Texcoco were highly encouraging, and late in 1997 another trial test was performed at the Cathedral's west atrium. The trial was especially relevant because it was necessary to have a reliable, well-established procedure to carry out the task beforehand.

Grout tests at the Cathedral

The grouting experiment began in December 1997 at the Cathedral's west atrium. From its results it was possible to deem this technique as suitable to solve problems caused by the accumulation of differential settlements in this monument. The presence of grouted mortar at the clay mass was verified by means of a sampling borehole that was also used to verify that the orientation of the mortar sheets at the treatment zone agreed with the theoretical estimates. Samples of the grouted clay were recovered with a 40 cm diameter pipe driven to a depth of 12.5 m (Figure 11.1). This boring was started from an open pit that crossed the foundation platform and the pre-Hispanic fills. During this experimental stage (13 November 1997 to 24 January 1998), 179.5 m³ of mortar were injected at the south-western zone of the Cathedral in 18 cores.

Additional studies

The project for hardening selectively the clay strata under the Cathedral was based on a comprehensive review of geotechnical conditions at the site including a reassessment of stratigraphical details and of the hydraulic conditions within the clay mass. In addition, seismic wave propagation velocities were determined with a seismic cone as well as the stress conditions of the soil mass through the use of a Marchetti dilatometer. Shear wave propagation velocities and the state of stresses measured in situ were determined before and after the grouting work. The effects of the grouting for improving the treated soil were established as reference targets for evaluation purposes, together with the topographic surveys and the levelling of deep benchmarks.

Drilling from the crypt level

Pneumatic and electric drilling rigs were used after being adapted to operate within the very narrow aisles of the crypts. Some of them were mounted on mobile bases to facilitate their transportation. In regard to drilling tools, pneumatic bottom hammers, triconic bits, simple

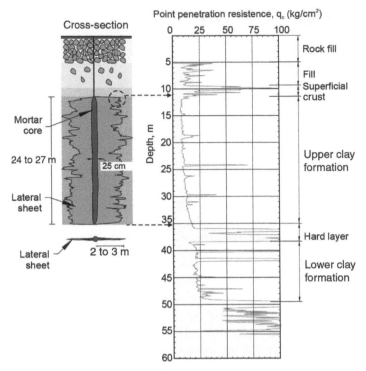

Point penetration resistence, q_c (kg/cm²)

Cross-section

Mortar core

24 to 27 m

25 cm

Lateral sheet

Lateral sheet

2 to 3 m

Depth, m

Rock fill

Fill
Superficial
crust

Upper clay
formation

Hard layer

Lower clay
formation

CPT sounding to define the injection depth

Sample of injected clay retrieved from a
depth of 12.7 m at the south-west tower

Figure 11.1 Verification of mortar injections during the trial injection at the south atrium.

Figure 11.2 Drilling and injection operations from the crypt level.

drag bits, and reamers were used at the depths where the cores would be grouted. The photograph in Figure 11.2 illustrates the drilling operations from the crypt level.

Drilling from the atrium

Heavy drilling rigs mounted on vehicles were used outdoors at both churches, as seen in Figure 11.3; only at certain stretches was ski-mounted equipment utilized. Borings were advanced with a procedure similar to that used at the crypts but with a somewhat larger diameter.

Figure 11.3 Drilling and injection operations from the south atrium.

Zones and percentages of grouting

The mortar grout injected to reduce the compressibility of the subsoil under the Cathedral is constituted of cement, bentonite, pumiceous sand and additives. Soils hardened by the grout are less deformable than the natural clay. Reductions of deformability depend on the stiffness of the mortar and on the percentage of grout injected (the ratio between the volume of mortar and the volume of soil to be improved). The borings to carry out the injection needed to cross the thickness of the rock fill, of the archaeological fills and of the superficial crust and then go through the clays of the upper clay formation that were grouted down to their contact with the first hard layer.

First stage of subsoil hardening

Figure 11.4 shows the areas where the subsoil of the upper clay formation under the Cathedral and the Sagrario was hardened. Grout percentages varied from 2 to 7% at the Cathedral and from 1 to 5% at the Sagrario. Deformable mortar was injected at 571 cores with their respective clusters of lateral sheets. There are 419 cores at the Cathedral, 111 at the Sagrario, and 41 at the museum building, all of them within the First Clay Formation. The cores were

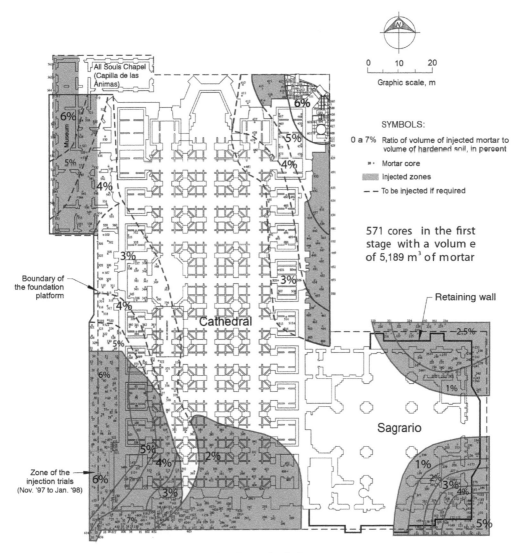

Subsoil grouting under the Cathedral . F irst stage: Sept. '98 to Sept. '99;
second stage: May to July 2000

Figure 11.4 Distribution of mortar injected into the deformable clays. First stage: September 1998 to September 1999; second stage: May to July 2000.

distributed taking due consideration of the distribution of compressibilities in the subsoil and was adjusted applying the observational method. The volume of injected mortar was 6,934 m^3.

Second stage of subsoil hardening

From 8 September 1998 to 4 June 1999 the southwestern part of the Cathedral was grouted as well as the northeast and southeast corners of the Sagrario (Figure 11.4). The southwest corner of the Cathedral was injected in two stages with a consumption of 50% of the total grout mix applied to each. Subsequently, from 7 June to 9 September 1999 the south zone was grouted; finally, in December 2000 the same thing was done at the northeast corner, although the volume of grout was of only a fraction of the total planned. The injection of grout sheets at the northeast part finished in December 2000 and in the museum a year later.

Chapter 12

Observed behaviour

Precision topographic levellings

These measurements were made at 246 control points distributed over the whole area covered by the monument. Column and pilaster plinths as well as chapel walls were levelled at the Cathedral and the Sagrario, including reference steel bolts embedded in the external walls. Points were located at the walls of the Capilla de las Ánimas (north west corner) and of the Bishopric (museum), at the grille bars of the atrium, and at the top of the deep benchmarks. These levellings were carried out every two weeks from October 1991 to the end of 1999; it was subsequently decided to schedule them monthly. A total of 215 levellings was made as of August 2000. Three levellings per year were done afterwards, until 2004. Two levellings were made in 2005 and one in 2007. The last levelling was made in August 2018.

The results are reported by means of contours that represent relative settlements accumulated since the beginning of the project in October 1991 and since the start of underexcavation in August 1993. The differential settlements produced every 28 days were also presented graphically. All the levellings are referred to an arbitrary reference point (the reference zero point) located near the apse at the intersection of axes C and 3, in the west side of the main altar. It was therefore possible to obtain the differential movements with respect to this point. All levellings are also referred to the deep benchmark installed at a depth of 100.4 m below the surface (BNP-100). The movements recorded represent the combination of those induced by regional subsidence and those produced by the effects of the corrective and preventive actions discussed and described before.

Efficacy of soil grouting

The efficacy of subsoil grouting can be evaluated by comparing the settlement rate of the Cathedral and the Sagrario before and after injecting mortars. The left-hand side of Figure 12.1 shows a picture of the initial behaviour expressed graphically by plotting the settlement rates observed between 7 January and 2 September, 1991. As seen in the figure, the central part of the Cathedral apparently emerged with respect to its northeast corner, at a rate of 16 mm/year, and with respect to the western bell tower at a rate of 12 mm/year. The maximum differential rate of 18 mm/year developed between the centre and the northeast corner. The Sagrario shows a maximum settlement rate at its southeast corner of 16 mm/year with respect to point C-3, located close to the apse of the Cathedral, and 20 mm between the southeast corner and the northwest columns.

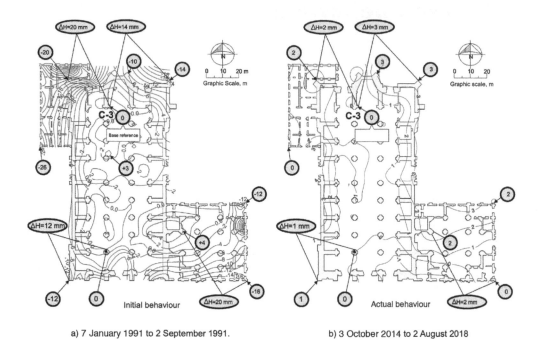

a) 7 January 1991 to 2 September 1991.　　　　b) 3 October 2014 to 2 August 2018

Figure 12.1 Results of differential levellings at the beginning of the project and during the 2014–2018 period.

The last levelling was performed in August 2018 and, as seen in Figure 12.1, differential settlements that accumulated up to that date followed the same general trend during previous years. That is, overall the Cathedral continued to sink close to uniformly. Settlement rates between point C3 and the southwest corner were very small, indeed negligible, over the 2014 to 2018 period. At the Sagrario Church, the rate at the northeast corner with respect to point C3 was 2 mm/year only. The results of that levelling also include the effects of the earthquake of 19 September 2017, an event with a magnitude $M_w = 7.1$ that hit the city from an epicentre located about 120 km south east of the city. Making reference to the data shown in Figure 12.1 in that the general settlement trends continued steadily with no apparent modification, it therefore follows that the earthquake did not affect settlement rates in a measurable way. It must be said that some ornamental sculptures in the façade toppled and fell over and that some minor cracking did occur as a consequence of the earthquake.

Chapter 13

Final remarks

The objectives for the geometrical correction of the Cathedral and of the adjacent Sagrario church were established on the basis of the experience previously gained while recovering the verticality of several other buildings. The method was adapted to a masonry structure and applied experimentally at the San Antonio Abad Church. Underexcavation at the Cathedral and the Sagrario started in August 1993. The preliminary goal was defined by Dr. Fernando López-Carmona; subsequently, this goal was modified in 1994 by Dr. Roberto Meli. In applying these two proposals sequentially, underexcavation actually satisfied both. Once underexcavation complied with the correction objectives established by the structural advisors of the project, the Advisory Technical Committee decided to conclude it in May 1998. Vertical corrective settlements after almost five years stabilized at a maximum of 88 cm.

The need to prevent the long-term effects of regional subsidence justified the application of mortar injections. The background information was taken from the case history of the Teatro Nacional, now the Palace of Fine Arts, that was reassessed applying basic soil mechanics concepts, unknown in 1906 to the initial builders of the theatre. The implementation of this method of subsoil hardening under the Cathedral was based on field and laboratory theoretical and experimental studies.

Evolution of differential settlements sustained by the Cathedral and the Sagrario up to August 2018 demonstrated that the mortar injection of the subsoil had beneficial effects on the behaviour of both churches. Settlement contours of at the level of the plane of plinths confirm that the historic settlement patterns were most favourably modified. The behaviour of both towers and of the reference plumb line corroborate the previous conclusions. Both underexcavation and mortar grouting were definitively efficient remedial methods.

The favourable results, confirmed by the comparison of the graphs presented in Figure 11.4 must be reassessed and reconsidered in subsequent topographical and structural observations that may even be used to device or implement other interventions to preserve the Cathedral and the Sagrario Church, if necessary. It had been assumed originally that underexcavation could be repeated periodically, possibly every 25 years, for the purpose of reducing future differential settlements. It was also envisioned that during that period perhaps a new corrective technique could be available to achieve additional benefits. In analyzing this situation, selective hardening of the subsoil emerged as a plausible possibility.

The data and analyses presented herein demonstrate that the subsoil hardening under the churches will become an alternative that, together with other complementary actions, will help preserve these monuments. Hardening has also the advantage of being a preventive method as opposed to underexcavation that represents a corrective action. The behaviour of both churches up to December 2007 suggests selective soil hardening is a long-term solution to the problems experienced by the Cathedral and the Sagrario. Further topographical level-lings performed in 2018 ratified this conclusion

Finally, it should be acknowledged that the geometrical correction and the subsoil hard-ening under the Cathedral and the Sagrario are actions that mitigate the harmful effects of differential settlements and they constitute examples of projects meticulously controlled through the observational method. This experience should lead the way and promote the development of practical research oriented towards the solution of many of the geotechnical problems still found in Mexico City.

References

Burland, J.B., Jamiolkowsky, M. & Viggiani, C. (2002) Preserving Pisa's treasure. *Civil Engineering Magazine*, March, 42–49.

Burland, J.B., Jamiolkowski, M. & Viggiani, C. (2013) The leaning Tower of Pisa. In: Bilotta, E., Flora, A. & Viggiani, C. (eds.) *Geotechnics and Heritage*, Taylor and Francis Group, London, ISBN 978-1-138-00054-4. pp. 207–227.

Jamiolkowski, M. (2001) The leaning Tower of Pisa: End of an Odyssey. *Proc. International Conference on Soil Mechanics and Foundation Engineering*, Istanbul.

Jamiolkowski, M., Lancellotta, R. & Pepe, C. (1993) Leaning Tower of Pisa: Updated information. *Proc. Third International Conference on Case Histories in Geotechnical Engineering*, Vol 2, St. Louis, Missouri. pp. 1319–1330.

Jamiolkowsky, M., Viggiani, C. & Burland, J. (1999) Geotechnical aspects. *Proc., Workshop on the Restoration of the Leaning Tower: Present Situation and Perspectives*, Preprints Volume, Pisa, July.

Lizzi, F. (1983) Preserving the original static scheme in the consolidation of old buildings. *IABSE Symposium, Strengthening of Building Structures, Final Report*, Vol 46, International Association for Bridge and Structural Engineering.

Mazari, M., Marsal, R.J. & Alberro, J. (1985) *The Settlements of the Aztec Great Temple Analysed by Soil Mechanics*. Mexican Society for Soil Mechanics, Mexico.

Meli, R. & Sánchez, R. (1995) Diagnóstico estructural, Chapter 4. In: *Catedral Metropolitana: corrección geométrica, informe técnico*. Asociación de amigos de la Catedral Metropolitana de México, A.C., Mexico City. pp. 141–188.

Moreno-Pecero, G. (1981) Refoundation research of Mexico's National Palace. *Proc. Xth International Conference on Soil Mech and Foundation Engineering*, Vol 2, Stockholm, Sweden. pp. 133–139.

Ovando-Shelley, E., Cuevas, A. & Santoyo, E. (1994) Assessment of the underexcavation technique for levelling structures in Mexico City: The San Antonio Abad case. *Proc. XIII Int Conf on Soil Mech and Foundation Engineering*, Vol 4, New Delhi, India. pp. 1461–1466.

Ovando-Shelley, E., Romo, M.P. & Ossa, A. (2007) The sinking of Mexico City: Its effects on soil properties and seismic response. *Soil Dynamics and Earthquake Engineering*, 27(4), 333–343.

Terracina, F. (1962) Foundations of the Tower of Pisa. *Geotechnique*, 12(3).

Archaeological aspects

Data for this appendix was provided by Álvaro Barrera, Archaeologist at the National Institute of Anthropology and History (Instituto Nacional de Antropología e Historia, INAH). Barrera conducted archaeological investigations at the site that provided valuable data and information that have expanded the knowledge about the Aztecs' religious precinct in central Mexico City.

Introduction

Most of the archaeological findings of the past few decades in Central Mexico City came to light as a result of excavations made for the construction of new buildings. Archaeological evidence has been randomly made public. Nonetheless, data collected during the development of these projects has contributed to gain more knowledge about the cultural development of the area. The Program of Urban Archaeology (PAU) was created in 1991 to associate archaeological field work in various parts of the city to studies made at the Aztec Religious precinct that spread around the Main Temple (Templo Mayor) and to those carried out at other neighbouring sites. Archaeological research in the area spreads further into the past, far beyond the Aztec period, and it implies analyses of the seven centuries of human settlements there.

The project for the Geometrical Correction of the Metropolitan Cathedral and of the adjacent Sagrario church was a major and unique opportunity for carrying out archaeological explorations at the southwestern part of the Aztec ceremonial precinct that was undoubtedly regarded as important by the Spanish builders to justify the erection of the most important shrine in the country in that part of the city. The excavation of 32 shafts and the execution of other civil engineering works that were carried out between the years 1991 to 1997 made it possible to dig into and recover the data hidden under these colonial buildings.

Summary of archaeological data

PAU was responsible for excavating 83 points inside the shafts, at the wells for driving piles, at test borings, and at other exploratory wells. Archaeological data was retrieved from all of them. Archaeological elements were recorded draughting plans, cross-sections, elevations and profiles that were backed by black and white photographs, colour slides, colour prints, and sometimes video films. Reports of all of the archaeological activities as well as daily, weekly, monthly and quarterly summaries were prepared. Field information was meticulously recorded in notebooks. Field and office activities were documented in logbooks. In addition, the supervising company was notified about the daily activities performed by all the members of the archaeological team. The sketches in Figure A.1 provide examples of archaeological interpretations of the structures found in two of the underexcavation shafts.

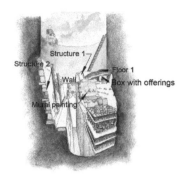
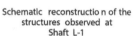
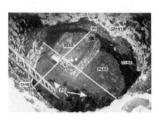
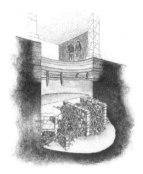

Structures uncovered at
Shaft L-1

Schematic reconstructio n of the
structures observed at
Shaft L-1

Findings at the core of a s tructure
found at Shaft L-7

Figure A. I Reconstruction of pre-Hispanic structures from archaeological data collected from the base
of two underexcavation shafts.

Archaeological findings

To have an overall idea of the amount of information gathered from the 32 shafts excavated, suffice is to mention that more than 30 sloping, straight and circular walls had been discovered, together with a large number of stair flights equivalent to 69,152 floor fragments and to 302 stratigraphic layers of cultural deposition.

Distribution of the structures uncovered

In the various drawings and scale models that archaeologists have presented to represent the possible distribution of the structures and monuments in the Aztec Ceremonial Centre, the temples are always surrounded by large plazas. However, as a result of this archaeological rescue work, this interpretation has been superseded because in certain cases the temples were very close to each other and occasionally, less than one metre apart.

Figure A.2 presents the preliminary interpretation of the pre-Hispanic building distribution at the site. The oldest era corresponds to the reigns of the Aztecs governors Chimalpopoca, Izcóatl, and partly of Moctezuma the first (1420–1447). The most recent buildings must have been erected by the year 1500 during the government of Ahuízotl. Construction layers exist between the stages referred to before and subsequent to them. However, for a better rendering of the distributions, only the two stages representative of the major architectural changes have been illustrated in the figure. Plan views with the location of each of them have been plotted. Only three buildings have been identified: the Teotlachco, that was the most important ball court of the Mexica ceremonial precinct, the Temple to Tonaituh, the Sun, and the Temple of Ehécatl Quetzalcóatl, God of the wind.

Archaeological data also evinced that the Aztecs buildings were also affected by differential settlements. The sketch in Figure A.3, for example, shows how they tried to correct inclinations and tilts in their temples.

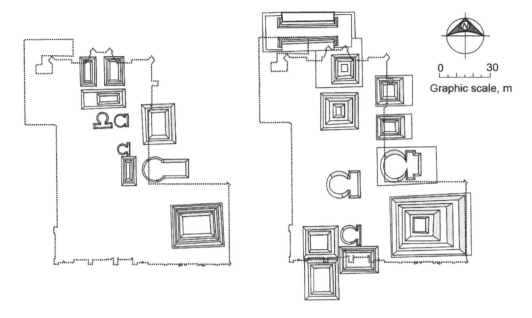

Graphic scale, m

First constructive stages
(1420-1477, Chimalpopoca, Izcóatl,
Moctezuma I)

Last constructive stage
(1500, Ahuizótl)

Figure A.2 Tentative distribution of pre-Hispanic structures under the Metropolitan Cathedral and the Sagrario Church.

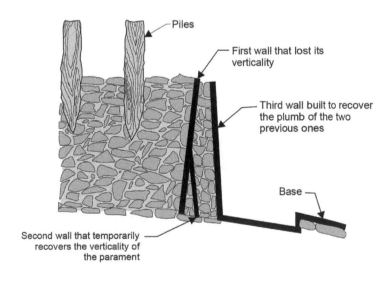

Piles

First wall that lost its verticality

Third wall built to recover the plumb of the two previous ones

Base

Second wall that temporarily recovers the verticality of the parament

Correction of loss of verticality at an Aztec
wall, as observed at Shaft L-4

Figure A.3 Corrections performed by the Aztecs to correct some of the effects of differential settlements on their structures.

Offerings

More than 20 offerings were found and logged, some of which were located inside and out of the remains of Aztec buildings, in stone or wooden boxes, on top and under floors and pavements, inside wicker containers or maguey leaves (American agave). Some of them contain only human bones; others, only maguey (agave) thorns and a few of them, miscellaneous objects (see Figure A.4).

The main contribution of this archaeological study to the knowledge of the Aztec civilization was the unveiling of the distribution of the structures found under the Cathedral, some of them built, as mentioned before, with spacings of less than a metre; they were subsequently joined together to integrate larger bodies. Prior to those explorations, it was commonly held that the structures were always surrounded by wide open spaces. The second contribution of this research was the rescue of valuable artefacts of ceramics, wood, and bone used in old times, before and after the Spanish occupation period. Several epochs have been recorded because in this area there have been human settlements for more than 650 years.

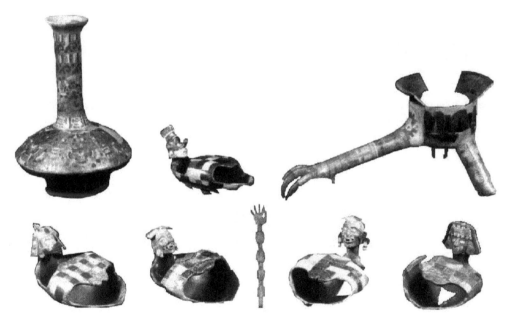

Figure A.4 Some of the miscellaneous objects found in one of the offerings found during the archaeological investigations.

General chronological account of the Cathedral and of the Sagrario Church

16th century

1521   Mexico-Tenochtitlán falls to the Spanish conquerors.
1524   Construction of the first church or Cortés' Cathedral starts.
1532   The church is granted the status of Cathedral.
1547   The status of Metropolitan Cathedral is assigned.
1555   The first dedication of the first Cathedral is celebrated.
1562   An attempt is made to build the Cathedral along an east-west main axis.
1573   The erection of the Cathedral along a north-south begins.
1581   The masonry platform for the foundation is completed.

17th century

1601   The first Cathedral is repaired.
1608   The Nochistongo outfall tunnel, north of Mexico City, is excavated.
1613   The foundation beams are finished 3.5 m above the level of the Main Square.
1622   Construction of the columns at the northern part is completed.
1623   The Sacristy is finished.
1625   The first Cathedral is demolished.
1629   The city is flooded and remains so until 1634.
1629   The construction of the Cathedral is delayed six years.
1635   The erection of the vaults and of the transept starts.
1637   Vaults for the chapels are completed.
1654   Columns of the southern part are finished (85 cm taller than those at the north).
1656   First dedication of the Cathedral.
1656   The erection of the main dome begins.
1664   The dome is completed.
1667   The vaults are finished; second dedication of the Cathedral.
1672   The main facade starts being sculpted; it is completed three years later.

18th century

1725   Jerónimo de Balbás completes the Main Altar (Altar de los Reyes).
1749   Construction of the Sagrario begins under Lorenzo Rodríguez (parish church of the Cathedral).
1768   The Sagrario is completed.

1780   José Damián Ortiz de Castro starts the construction of the bell towers.
1789   The Nochistongo tunnel is converted into an open channel.
1792   The bell towers are completed.
1793   Manuel Tolsá starts working at the Cathedral.
1795   The erection of the Seminary building starts.
1796   The Sagrario catches fire.

19th century

1800   The Seminary is completed.
1813   The final construction works at the Cathedral are finished.
1860   Javier Cavallari executes the first topographic levelling of the western bell tower.
1870–76   Juan Cadena repairs the Cathedral.
1884   An earthquake damages the west side of the Cathedral.
1885   Ramón Agea starts the repair of the damages suffered by the west side of the Cathedral.
1894   A. Torres Torrija and R. Gayol evaluate the damages caused by the 1884 earthquake.
1898   Ramón Agea completes the repair works of the west side of the Cathedral.

20th century

1906   Luis G. Olvera starts the restoration of the arches and vaults.
1907   Roberto Gayol carries out topographic levellings.
1910   An earthquake shakes Mexico City.
1925   G. Olvera completes the repairs of arches and vaults.
1927   M. Ortiz Monasterio and M. Cortina start their reconstruction schemes.
1929   Manuel Ortiz Monasterio delivers his project to rehabilitate the foundation.
1938   The Seminary building is demolished.
1942   Antonio Muñoz is sworn as Director of Conservation.
1943   Lorenzo de la Hidalga's main altar is demolished.
1957   An earthquake occurs in Mexico City that knocks down the Angel of the monument to the Independence.
1959   Antonio Muñoz quits as Director of Conservation.
1967   The Altar del Perdón catches fire.
1972   Subsoil studies are started; the remains of the first Cathedral are uncovered.
1972   Manuel González Flores delivers his underpinning project.
1974–76   Control piles are driven to underpin the Cathedral; grouting of the vaults takes place.
1985   Mexico City experiences a devastating earthquake on 19 September. The Cathedral suffers minor damages.
1989   In April water seeps through the vaults; authorities are concerned.
1989   Geotechnical studies are carried out from June to September.
1990–91   Between May 1990 and January 1991 underexcavation begins.
1991–93   Between October 1991 and December 1993 the first 30 shafts are excavated.
1993–98   Between August 1993 and June 1998, underexcavation continues.
1997–98   Mortar injection is evaluated experimentally.
1998–99   Between September 1998 and September 1999 the first stage of the subsoil injection is carried out.

21st century

2000   The second stage of the subsoil injection takes place.
2017   Another large earthquake hits Mexico City. Minor damage to the Cathedral.

Master builders, architects and engineers

A compilation of historic documents on the Cathedral and the Sagrario was carried out to find out all the significant construction aspects and all the repairs made to the Cathedral since the 17th century. The decisions and reasoning followed by the master builders, architects and engineers are so wide in scope that their discussion or even their description falls outside the aims of this publication. The names and the works performed by master builders, masons, architects, engineers, conservationists and others that have contributed to the construction and to the preservation of these two monuments is provided next

Master builders, masons and architects for the construction

1   Of the first Cathedral built by order of Hernán Cortés at the present south atrium:

- Maese Martín de Sepúlveda, builder of the first Cathedral in 1524.
- Alonso Arias, restorer of the old Cathedral from 1601 to 1602.

2   Of the Cathedral and the Sagrario beginning in 1573 until their completion in 1813:

- Claudio de Arciniega, master builder who conceived the Cathedral.
- Juan Miguel de Agüero, master builder who succeeded Arciniega.
- Alonso Pérez de Castañeda, major mason of the construction between 1573 and 1615.
- Juan Serrano and Melchor Pérez de Soto, continued the building in 1655.
- Juan Gómez de Trasmonte, major mason who erected the vaults and the transept in 1630.
- Luis Gómez de Trasmonte and Rodrigo Díaz de Aguilera, builders of the dome in 1656.
- Cristóbal de Medina Vargas, major mason in 1684.
- Felipe de Roa and Antonio de Roa, successive major masons between 1699 and 1709.
- Jerónimo de Balbás, creator of the altarpieces "de los Reyes", "Ciprés", and "del Perdón".
- Pedro de Arrieta, builder in 1725 of the Seminary, demolished in 1936.
- Manuel de Álvarez, reviewer of the design of the presbytery made by Jerónimo de Balbás.

- Lorenzo Rodríguez, designer and builder of the Sagrario between 1749 and 1768.
- Francisco Antonio Guerrero y Torres, follower of the work of Lorenzo Rodríguez.
- José Damián Ortiz de Castro, started the construction of the western bell tower in 1787.
- Francisco Ortiz de Castro, continued the works at the Cathedral after the death of his brother.
- Ignacio Castera, collaborator of Ortiz de Castro in the design of the bell towers.
- Manuel Tolsá, harmonized the Cathedral from 1793 to 1813.

Architects and engineers for the repair of damages

1   Appointed by the Academy of Fine Arts (Academia de Bellas Artes):

- Javier Cavallari, executed the first topographic levelling of the western bell tower in 1860.
- Juan Cadena, repaired the Cathedral between 1870 and 1876.

2   Appointed by the Secretary of State for the Treasury (1894):

- The Cathedral is inspected by orders of José Yves Limantour. The committee formed by Antonio Torres Torrija, Mateo Plowers and Roberto Gayol assessed the damages caused by the 1884 earthquake.
- Luis G. Anzorena and Antonio Torres Torrija, appraised the condition of the Cathedral.
- Mauricio M. Campos and Antonio Rivas Mercado, described the damages to the Cathedral in 1905.
- Roberto Gayol y Soto, carried out topographic levellings of the Cathedral in 1895 and 1907.
- Ramón Agea, responsible for the repair works of the west side between 1885 and 1898.

3   Appointed by the Secretary of State for Education (Secretaría de Educación):

- The condition of the Cathedral is assessed by order of José Vasconcelos.
- Luis G. Olvera supervised repairs to dome and vaults; 14 arches with steel bands and a reinforcing arch with steel bars were added (1906–25).

4   Appointed by the Technical Committee for the Conservation of the Cathedral (Comisión Técnica y Conservación de la Catedral, 1927):

- Daniel García and Luis McGregor appraised the condition of the Cathedral.
- Manuel Cortina García, Manuel Ortiz Monasterio and Roberto Gayol developed the project for the rehabilitation of the Cathedral in 1929.

5   Appointed by the Diocesan Committee of Order and Decorum (Comisión Diocesana de Orden y Decoro, 1937):

- Nicolás Mariscal y Piña, director of the Cathedral works in 1940 and 1941.
- Antonio Muñoz García, architect who was Director of Preservation of the Cathedral from 1942 to 1959.

- Alberto J. Flores and Manuel González Flores, appraisers of the damages caused by the fire of 1967.

6   Appointed by Laboratorio ICA:

(Note: This company performed a topographic survey of the Cathedral)

- Raúl J. Marsal and Marcos Mazari, analyzed differential settlements (1953–55).

Technicians during underpinning with control piles

Appointed by the Secretary of State for National Patrimony (Secretaría del Patrimonio Naciona, SPN, 1972):

- Vicente Medel, Director of Monuments for SPN.
- Pedro Moctezuma Díaz Infante, Assistant Secretary for SPN and project director.
- Manuel González Flores, designer and contractor of the underpinning project with control piles.
- Vicente Guerrero y Gama, evaluated the loads transmitted by the Cathedral to the foundation.
- Ernesto Martínez Parker, responsible for the geotechnical study.
- Jaime Ortiz Lajous, project coordinator.
- Agustín Salgado, resident of the works executed from 1974 to 1976.

Engineers and architects for underexcavation and selective soil hardening

The project was carried out under the Director General of Sites and Monuments of Cultural Patrimony (Dirección General de Sitios y Monumentos del Patrimonio Cultural, 1989–99), and under the General Coordination of Special Projects (Coordinación General de Obras Especiales, 1999–2000).

Appointed by the Secretary of State for Urban Development and Ecology (Secretaría de Desarrollo Urbano y Ecología, 1989), the Secretary of State for Social Development (Secretaría de Desarrollo Social, 1992), the Secretary of State for Public Education (Secretaría de Educación Pública, 1994), and the National Council for Arts and Culture (Consejo Nacional para la Cultura y las Artes, 1997):

- Sergio Zaldívar Guerra, Project Director (1989–2000)
- Xavier Cortés Rocha, Project Director (2000–06).
- Fernando López Carmona, Roberto Meli Piralla, and Hilario Prieto, members of the Technical Committee and structural engineers for the project.
- Enrique Tamez and Enrique Santoyo, members of the Technical Committee and geotechnical engineers for the project.
- Fernando Pineda, member of the Technical Committee.
- Jorge Díaz Padilla, secretary of the Technical Committee.
- Efraín Ovando-Shelley and Roberto Sánchez, geotechnical and structural advisors, respectively.

Appendix D

Geotechnical dissemination of the project

The objectives and the progress of the geotechnical works that were carried out at the Cathedral have been published in journals, the proceedings of many technical meetings, lectures, discussion sessions, symposia and seminars in Mexico and abroad. Special mention should be made of the special lecture delivered on occasion of the XIV International Conference of Soil Mechanics and Foundation Engineering at Hamburg in 1997 and during the 2004 Skempton Memorial Conference in London. The papers and reports dealing with the geotechnical aspects of the project are listed below in chronological sequence:

Ovando-Shelley, E. & Manzanilla, L. (1997) Archaeological interpretation of geotechnical sounding in the Metropolitan Cathedral, Mexico City. *Archaeometry*, 39(1), 221–235, México.

Ovando-Shelley, E. & Santoyo, E. (2001) Underexcavation for levelling buildings in México City. *Journal of Architectural Engineering, ASCE*, 7(3), 61–70. Conference, Institution of Civil Engineers, Londres, Vol 2. pp. 1155–1168.

Ovando-Shelley, E. y Santoyo, E. (2003a) Paralelismo entre la torre de Pisa y la Catedral Metropolitana de la ciudad de México (primera parte). *Ingeniería Civil*. Colegio de Ingenieros Civiles, México, año LIII, No. 409, 22–23 y 30–34.

Ovando-Shelley, E. y Santoyo, E. (2003b). Paralelismo entre la torre de Pisa y la Catedral Metropolitana de la ciuad de México (segunda parte). *Ingeniería Civil*. Colegio de Ingenieros Civiles, México, año LIII, No. 410, 5–15 y 24–25.

Ovando-Shelley, E. & Santoyo, E. (2013) Contributions of geotechnical engineering for the preservation of the Metropolitan Cathedral and the Sagrario Church in Mexico City. Capítulo en: Bilotta, E., Flora, A., Lirer, S. & Viggiani, C. (eds.) *Geotechnics and Heritage*, Taylor & Francis Group, London, ISBN 978-1-00054-4. pp. 153–178.

Ovando-Shelley, E. & Takahashi, V. (1996) Impact of regional subsidence and changing soil properties on the preservation of architectural monuments in Mexico City. *Proc. Int. Symposium on Geotechnical Engineering for the Preservation of Historical Sites*, Preprints Volume, University of Naples. pp. 633–641.

Ovando-Shelley, E. & Tamez, E. (1998) Geometrical correction of Mexico City's Metropolitan Cathedral by means of underexcavation. *Felsbau, Rock and Soil Engineering*, 16(6), 450–458, Austria.

Ovando-Shelley, E., Cuevas, A. & Santoyo, E. (1994) Assessment of the underexcavation technique for leveling structures in Mexico City: The San Antonio Abad case. *Proc. XIII Int. Conf. on Soil Mech. and Foundation Engineering*, Vol 4, Nueva Delhi, India, ed. Balkema. pp. 1461–1466.

Ovando-Shelley, E., Tamez, E. & Santoyo, E. (1996) Geotechnical aspects for underexcavating Mexico City's Metropolitan Cathedral: Main achievements after three years. *Proc. Int. Symposium on Geotechnical Engineering for the Preservation of Historical Sites*, Preprints Volume, University of Naples. pp. 456–465.

Ovando-Shelley, E., Tamez, E. & Santoyo, E. (1997) Options for correcting differential settlements in Mexico City's Metropolitan Cathedral. *Proc. XIVth Int. Conf. on Soil Mechanics and Foundation Engineering*, Hamburgo.

Ovando-Shelley, E., Romo, M.P. & Ossa, A. (2007) The sinking of Mexico City: Its effects on soil properties and seismic response. *Soil Dynamics and Earthquake Engineering*, 27(4), 333–343.

Ovando-Shelley, E., Pinto Oliveira, M.O., Santoyo Villa, E. & Hérnández, V. (2008) Mexico City: Geotechnical concerns in the preservation of monuments. *International Journal of Architectural Heritage*, 2, 1–23.

Ovando-Shelley, E., Ossa, A. & Santoyo, E. (2013) Effects of regional subsidence and earthquakes on architectural monuments in Mexico City. *Boletín de la Sociedad Geológica Mexicana*, 65(1), 157–167.

Ovando-Shelley, E., Santoyo-Villa, E. & Pinto de Oliveira, M.A. (2013) Intervention techniques. Keynote lecture in: Bilotta, E., Flora, A., Lirer, S. & Viggiani, C. (eds.) *Geotechnical Preservation of Monuments and Historic Sites*, University of Naples Federico II, Napoli, Italy, ISBN 978-1-138-00055-1. pp. 75–91.

Ovando-Shelley, E, Santoyo-villa, E. & Hernández, J. (2015) Mexico City's Metropolitan Cathedral and Sagrario Church thirteen years after underexcavation and soil hardening. *International Journal of Architectural Heritage: Conservation, Analysis, and Restoration*. DOI:10.1080/15583058.2015.1113331.

Pinto Oliveira, M., Ovando-Shelley, E. y Santoyo Villa, E. (2006) La ingeniería de cimentaciones en algunos monumentos arquitectónicos de Venecia y la ciudad de México. Memorias, XXIII Reunión Nacional de Mecánica de Suelos, Sociedad Mexicana de Mecánica de Suelos, Tuxtla Gutiérrez, Chiapas, Vol 1, 275–284.

Santoyo, E. y Ovando-Shelley, E. (1995) La Torre de Pisa y la Catedral de México: semejanzas y diferencias. *Memorias X Congreso Panamericano de Mecánica de Suelos e Ingeniería de Cimentaciones*, Vol 4, Sociedad Mexicana de Mecánica de Suelos.

Santoyo, E. y Ovando-Shelley, E. (1998) Termina la subexcavación de la Catedral Metropolitana. *Vector de la Ingeniería Civil* (14), 18–29, México.

Santoyo, E. & Ovando-Shelley, E. (2001) Injected mortars to reduce the compressibility of the subsoil in Mexico City's Metropolitan Cathedral. *Proc., XVth International Conference on Soil Mechanics and Geotechnical Engineering*, Vol 2, Estambul, Turquía. pp. 1541–1544.

Santoyo, E. & Ovando-Shelley, E. (2002) Underexcavation at the Tower of Pisa and at Mexico City's Metropolitan Catedral. *Proc. International Workshop, ISSMGE: Technical Committee TC36Foundation Engineering in Difficult*, Soft Soil Conditions, CD edition, Mexico City.

Santoyo, E. & Ovando-Shelley, E. (2003a). Behaviour of Mexico City's Metropolitan Catedral after underexcavation and soil hardening. *Proc. Int. Conf. Dedicated to the tercentenary o Saint Petersburg, Reconstruction of historical cities and Geotechnical Engineering*, Vol 1, San Petersburgo, Rusia, ASV Publishers, Moscú. pp. 255–266.

Santoyo, E. & Ovando-Shelley, E. (2003b) Cement injection in Mexico City for levelling buildings. Cap 12 en: Passado, presente e futuro dos edificios da orla marítima de Santos. Sao Paulo, Asociación Brasileña de Mecánica de Suelos. pp. 141–178.

Santoyo, E. y Ovando-Shelley, E. (2004a) Catedral y Sagrario de la ciudad de México. In: *Corrección geométrica y endurecimiento del subsuelo*. Consejo Nacional para la Cultura y las Artes, Dirección de Sitios y Monumentos del Patrimonio Cultural. En prensa, México.

Santoyo, E. & Ovando-Shelley, E. (2004b) Geotechnical considerations for hardening the subsoil in Mexico City's Metropolitan Catedral. *Proc. Skempton Memorial Conference*, Vol 2, Institution of Civil Engineers, London. pp. 1155–1168.

Santoyo, E. y Segovia, J.A. (1995a) Estructuras, recimentación y renivelación. *Ingeniería Civil*, n. 314, Colegio de Ingenieros Civiles de México. pp. 9–21.

Santoyo, E. y Segovia, J.A. (1995b) *Recimentación y renivelación de estructuras y monumentos*. Publicación de TGC Geotecnia, SA.

Santoyo, E., Ovando-Shelley, E., Guzmán, X., Cuanalo, O. y De la Torre, O. (1998) Palacio de Bellas Artes. In: *Campañas de inyección del subsuelo*. Publicación de TGC Geotecnia, Mexico City.

Santoyo, E., Ovando-Shelley, E. y Segovia, J. (1999) Evolución de las cimentaciones en la ciudad de México (primera parte). *Vector de la Ingeniería Civil* (21), 16–25.

Santoyo, E., Ovando-Shelley, E. y Segovia, J. (1999a) Evolución de las cimentaciones en la ciudad de México (segunda parte). *Vector de la Ingeniería Civil* (22), 9–13.

Santoyo, E., Ovando-Shelley, E. y Segovia, J.A. (1999b) Evolución de las cimentaciones de edifica-ciones en la ciudad de México. Ingeniería Civil, Centro de Estudios y Experimentación de Obras Públicas, Ministerio de Fomento, España, 116. pp. 63–81.

Santoyo Villa, E., Ovando-Shelley, E. y Pinto Oliveira, M. (2010) Principios para subecavar la cat-edral y el sagrario metropolitanos. En: Fernando López Carmona. 50 años de enseñanza, X Guzmán Urbiola, A. Hernández y I. San Martín, Compiladores. México, DF: Facultad de Arquitectura, Uni-versidad Nacional Autónoma de México, ISBN-978-607-02-0967-3. pp 138–155.

Tamez, E., Santoyo, E. y Cuevas, A. (1989) Estudio de las cimentaciones de la Catedral y el Sagrario de la Ciudad de México, Informe técnico preparado por TGC Geotecnia para la Secretaría de Desar-rollo Urbano y Ecología.

Tamez, E., Santoyo, E. y Cuevas, A. (1992a) La Catedral Metropolitana y el Sagrario de la Ciudad de México. In: Marsal, R.J., Ovando-Shelley, E. y Auvinet, G. (eds.) *Corrección al comportamiento de sus cimentaciones*, Sociedad Mexicana de Mecánica de Suelos, Mexico City.

Tamez, E., Santoyo, E. y Cuevas, A. (1992b) Rescate 1ª parte: Enfrentando el hundimiento de la Catedral; "Rescate 2ª parte: Subexcavación de las zonas duras; "Rescate 3ª parte: Excavación de lumbreras. Ingeniería Civil, núms. 284, 285 y 286, Colegio de Ingenieros Civiles de México.

Tamez, E., Santoyo, E. y Ovando-Shelley, E. (1995a) Diagnóstico y proyecto geotécnico. Catedral Metropolitana: Corrección geométrica, cap. 2, Asociación de Amigos de la Catedral Metropolitana de México.

Tamez, E., Santoyo, E. y Ovando-Shelley, E. (1995b) Procedimiento de subexcavación. Catedral Met-ropolitana: Corrección geométrica, cap. 6, Asociación de Amigos de la Catedral Metropolitana de México.

Tamez, E., Ovando-Shelley, E. & Santoyo, E. (1997) Underexcavation of the Metropolitan Cathedral in Mexico City. *Proc. XIVth Int. Conf. on Soil Mechanics and Foundation Engineering*, Hamburgo.

Index

Note: *Italic* page references indicate figures.

Built Heritage and Geotechnics

Milton Keynes UK
Ingram Content Group UK Ltd.
UKHW050440111024
449327UK00039B/18